AGRICULTURE ISSUES AND POLICIES

GRAZING ECOLOGY

VEGETATION AND SOIL IMPACT

AGRICULTURE ISSUES AND POLICIES

Additional books in this series can be found on Nova's website under the Series tab.

Additional e-books in this series can be found on Nova's website under the e-books tab.

AGRICULTURE ISSUES AND POLICIES

GRAZING ECOLOGY

VEGETATION AND SOIL IMPACT

JOSÉ RAMÓN ARÉVALO
EDITOR

Nova Science Publishers, Inc.
New York

For permission to use material from this book please contact us:
Telephone 631-231-7269; Fax 631-231-8175
Web Site: http://www.novapublishers.com

Library of Congress Cataloging-in-Publication Data

Grazing ecology : vegetation and soil impact / editor: Josi Ramsn Arivalo.
p. cm.
Includes index.
ISBN 978-1-62100-436-3 (hardcover)
1. Grazing--Management. 2. Pasture ecology. I. Arivalo, Josi Ramsn.
SF85.G725 2012
333.74--dc23
2011034000

Published by Nova Science Publishers, Inc. † New York

CONTENTS

PREFACE

Grazing management is one of the most important management activities worldwide, and it has been a traditional sustainable land use in many areas of the planet. This book demonstrates the poor extrapolation capacity that grazing studies have due to site-dependent effects and sensitivity to changes in weather conditions and soil characteristics. Therefore, researchers should be cautious when suggesting management recommendations, such as removing all grazing animals or reducing grazing activities. A highly variable ecological force such as grazing, with important environmental and social implications, should be analyzed deeply before extrapolated results or suggesting management practices.

Chapter 1 - Goat grazing management is one of the most important management activities worldwide, and it has been a traditional sustainable land use in many areas of the planet, requiring correct grazing management to maintain species composition and high diversity values of the plant communities in grazed lands. In spite of this traditional and sustainable use of goat grazing, the effects can cause remarkable and significant variation in species composition.

Grazing effects on species richness and species composition are contradictory when comparing studies, and the authors can find research concluding that herbivores can enhance, have weak effects, or even have a negative impact on plant diversity and richness, as well as different effects on soil nutrients. The lack of consistent responses among studies has been attributed to factors such as the evolutionary history of grazing, productivity gradients, or grazing intensity. The high context-dependence of grazing effects on ecosystems highlights the need to carry out studies that allow us to identify the intrinsic dynamics of the pastures in each region.

Chapter 2 - In the Indian Thar desert, different type of grazing practices is followed since long. Scientist from Central Arid Research Institute (CAZRI), Jodhpur has conducted several experiments on grazing and its impact on livestock, ecosystem and environment. It has been found that the ever increasing trend in the livestock and human population has put heavy grazing pressure and continuous decrease in the area of land available for grazing. Due to these factors, around 60% of grazing lands are in poor condition, 14% in fair, 13% in good and 2–3% in excellent state. Low and erratic rains and frequent drought restricts the vegetative and self seeding period of natural vegetation. The different conventional grazing experiment has provided useful information about the impact of grazing on rangelands of the Indian Thar desert. It has been observed that continuous grazing led into reduction in the density, basal cover and forage production of many palatable perennial species and increase of unpalatable species. The rotational and deferred rotational grazing systems are superior over the continuous controlled grazing. It has been observed these systems provide rest to the perennial species, which can assume their growth and vigor and produces high biomass and also these systems provide better opportunities to the grazing animals for utilizing animals and those species which become less palatable after maturity.

While, taking decisions on stocking rate of animals in rangeland it also should be kept in mind that the rangeland should not be overgrazed. Because, experimental studies have shown that overgrazing may lead to 3 to 5 times higher soil loss than controlled grazing due to wind erosion. Moreover, these eroded soils pose multifaceted problems in environment e.g. heavy atmospheric load of health hazardous fine dusts, loss of soil carbon and nutrients etc.

Therefore, grazing resources of the Indian Thar desert are to be managed in a sustainable manner specifically under the climate change scenarios and under increasing trend of grazing pressure in rangelands. Here, the authors have reviewed the existing grazing resources of the Indian Thar desert, livestock-pasture relationship under different grazing systems of the region, impact of different grazing systems on ecosystem and environment. This will help to formulate suitable rangeland policies in the Indian Thar desert with an aim to control the grazing pressure and to obtain sustained primary and secondary productivity.

Chapter 3 - Ungulate grazing in the Australian Alps commenced in the mid-19th Century. Prior to that, there were no large grazing animals in the alpine zone. Ecological research into the effects of domestic grazing commenced in the 1940s following concern about the effect of grazing animals on soil stability and a proposed hydro-electricity scheme. Grazing was found to increase the cover of bare ground and the risk of soil erosion. The creation of bare ground by grazing animals in grass-dominated vegetation facilitated the establishment of shrubs. In the absence of grazing, the cover of palatable forbs and shrubs increased. Cattle have had no influence on tall shrublands where there are few palatable species but they may affect the structure and function of wetland vegetation through trampling. Following a landscape scale fire in 2003, cattle were found to have had no effect on what vegetation was and was not burnt in the treeless country of the Alps. That is, fuel reduction using grazing was ineffective in fire mitigation.

Chapter 4 - On an oceanic island, the positive relationship between livestock management and biodiversity conservation is not as clear as on the mainland. In the sensitive island's ecosystems, introduced herbivores have been considered to cause devastating effects on vegetation, and be responsible for overgrazing, ecosystem degradation, and biodiversity loss. Eradication of introduced herbivores has been proposed as a conservation tool in some islands. However, this proposal needs to be carefully considered, since there are complex and established ecological interactions between native and introduced species.

The goal of this chapter is to discuss the livestock management for biodiversity conservation, with special attention on the Canary Islands (Spain) as an example. The authors present a review of the research done on this topic in the Canaries, looking for general conclusions and explanations of the conflicting results. Much of the research is restricted in area, however, consistent works are presented that can help us to understand the answer of the environment to grazing impact on these islands.

Although some negative impacts of grazing were revealed, in general they have been restricted to some specific areas or species. In those cases where goat grazing is threatening endemic and endangered populations, control of goat densities, grazing restrictions and use of fences should be encourage, however drastic measures as eradication of goats are not ecologically or socially feasible. On the other hand, grazing of small goat herds on traditionally managed ecosystems seems to be a sustainable activity, which removal have been related with a reduction in the number of species, changes in species composition and also an increase in exotic species.

Chapter 5 - Grazing and its interaction with other environmental variables generate temporal and spatial variations in plant communities. In the Patagonian steppe in southern Chile, two conditions were analyzed: a grazed active cattle ranch and a non-grazed area protected from domestic cattle during a 10-year period. The aim was to determine how the exclusion from domestic grazing affects species diversity and the replacement grasses of *Festuca gracillima*, with dwarf shrublands of *Empetrum rubrum*. Species richness, foliage coverage, and percent of bare soil at 1m^2 and 1000 m^2 scales were compared using MANOVA, to test possible differences due to grazing effects and edaphic conditions. Multivariate analyses (Jaccard's similarity coefficients and Detrended Correspondence Analysis) were used to test for community differences.

The results show that the *F. gracillima* grasses had higher species richness in non-grazed sites compared to the shrublands. At the 1000 m^2 scale, no significant difference in exotic species richness was found between the non-grazed and the grazed sites. Soil fertility was significantly higher in grasses with respect to shrublands at the 1000 m^2 scale. Jaccard's Index indicated a 49 % of floristic similarity between grazed and non-grazed grasses at the 1000 m^2 scale. The DCA at the 1 m^2 scale with presence/absence data indicated a continuous gradient with respect to the floristic composition between grass and shrublands species, independent of the grazing factor. A similar response was observed at the 1000 m^2 scale.

This research concludes that protection from grazing increased native species richness at the 1 and 1000 m^2 scales in only one decade. These rapid changes are most probably caused by the release from excessive grazing, which is threatening the conservation of plant species in the Austral Patagonian steppe.

Chapter 6 - Yellow starthistle (*Centaurea solstitialis*) is the most widely distributed invasive weed in California. It has infested an estimated 14.3 million acres and is present in 56 of the 58 counties. Yellow starthistle infestations cost the State of California millions of dollars in losses and impacts both natural and agricultural systems. Grazing is one of the available methods for the control of yellow starthistle and is already in use by private landowners and several land management agencies in California. However, starthistle produces sesquiterpene lactones that may cause adverse effects to large herbivores ranging from lethal to digestive discomfort. In addition the sesquiterpene lactones present in yellow starthistle can accumulate in animal tissues or secreted in milk of livestock used to control this weed. The presence of sesquiterpene lactones in yellow starthistle could hinder the implementation of grazing management programs to control this weed.

This chapter comprises a combination of field trials and detailed laboratory work aimed at determining the seasonal variability of nutritional and anti-nutritional characteristics of yellow starthistle and the effects of ingestion of yellow starthistle in sheep. The following conclusions were drawn from this chapter:

1. Yellow starthistle has good nutritional value for sheep. Feed intake and digestibility of yellow starthistle meet the nutritional requirements for maintenance for sheep.
2. The sesquiterpene lactones present in yellow starhistle do not have anti-quality properties and do not affect the health of sheep grazing this plant.

Based on the information gathered in this chapter, sheep grazing can be an effective tool in the control of yellow starthistle without deleterious effects to the health and well-being of the sheep.

Chapter 7 - Overgrazing has been regarded as a main cause for land degradation in Inner Mongolia grassland because of the increase in population and shift in the socio-economic system in recent years. However, given the vital importance for the production of livestock and the environmental changes, fundamental understanding on topsoil sensitivity to grazing is lacking. Based on MAGIM (Matter fluxes in grasslands of Inner Mongolia as influenced by stocking rate) project, intensive observations, analyses, and modeling of environmental changes responding to grazing in Inner Mongolia grassland were conducted. In this chapter, the authors concentrated on the spatio-temporal variation of soil properties (e.g., soil moisture) in general and modeling the effects of grazing management (i.e., five grazing plots including, ungrazed since 1979=UG79, ungrazed since 1999=UG99, moderately grazed=MG, continuously grazed=CG, and heavily grazed=HG) on water flux in particular. Results showed that grazing treatment reduced the topsoil moisture contents (0-6 cm) and homogenized soil moisture spatial patterns which were controlled by soil and plant properties, especially soil physical properties. Multivariate geostatistics further revealed scale-dependent correlation for these controlling parameters depending on treatment. The soil and plant properties strongly controlled the variation of soil moisture for UG99 at short-scale (45 m), and for CG and HG at long-scale (90 m), however, weakly controlled the variation of soil moisture for UG79 and MG. Therefore, the authors suggest that pasture management strongly modified soil moisture patterns, which should be considered in the hydrological models at multiple spatial scales. Based on soil pit scaled monitoring data, the authors further found seasonal change of the soil moisture was most significant in the topsoil (0-10 cm) and was less pronounced in the deeper soil. Soil water storage decreased with increasing grazing intensity, which is linked to the grazing-induced soil structure change. Moreover, we use HYDRUS-1D model to parameterize this grazing-induced shifts of soil pore function. The model was well verified with the measured soil moisture values. The modeling results showed no significant difference on water budget components between the two ungrazed sites and moderate grazing, while heavy grazing significantly decreased interception from 17 to 7 mm and transpiration from 121 to 74 mm, and increased evaporation from 88 to 128 mm. The authors conclude that intensive grazing in our studied area deteriorated soil functions and reduced plant available water, and consequently reduced grassland productivity and enhanced the risks for wind and water erosion.

Chapter 8 - Over the last two decades, agri-environment schemes have been developed by member states of the European Union. Recent assessments reveal the lack of success of these schemes in stopping biodiversity loss. While initial schemes focused on wildlife-friendly practices at field scale, the agro-landscape scale is increasingly considered as the relevant scale of analysis and intervention. Recent studies argue that efficient schemes require modifying both management practices at field scale and their spatial arrangements at landscape scale, in order to increase the spatial heterogeneity of agro-landscapes. To date however, only a few studies quantified the relationships between agricultural production and biodiversity conservation in farmlands. The objective of this chapter is to asses, at different spatial and temporal scales the trade-offs between production and bird conservation in a grassland agroecosystem used for meat cattle breeding. The research methodology is based on several models, built within the mathematical framework of the viability theory. These models formalize the interactions between grazing and mowing practices and the ecological dynamics of two grassland bird species at three nested spatial scales (field, farm and landscape). The five major findings of this thesis are as follows: at field scale (1) The timing and intensity of grazing are the major drivers of the trade-off between production and bird conservation, (2) the best ecological performances are obtained for intermediate levels of productive performances, (3) the inter-annual variations of management practices, made possible by result-oriented agri-environment schemes, improve ecological performances while at the same time increase grassland management flexibility, (4) at farm scale, the proportion of management regimes is a major driver of the trade-off between production and conservation, but bird conservation always implies a production cost, (5) at landscape scale, the spatial arrangement of different land uses improves the trade-off between production and conservation. Overall, these results reveal the different drivers of the production / conservation trade-offs at different scales. At each scale, the interactions between the different management regimes are the basis of the trade-off mechanisms between production and conservation. The importance of the spatial arrangement of management regimes suggests that the conciliation between production and conservation could strongly benefit from increased coordination among farms.

Chapter 9 - The growth of the world population demands more food production. This raises the need to conserve the natural resources and the environment. In the search to achieve this goal and with the emergence of new technologies, the production in Argentina has changed accordingly. Most of the croplands in Argentina are currently managed under continuous cropping (CC), which is based on the extensive use of zero-till (ZT) farming, application of herbicides and prevalence of soybean [*Glycine max* (L.) Merr.] within crop rotations. Although this production system is economically sound, at least on a short-term basis, it produces few agricultural products. The integration of crop and livestock (ICL) production is an interesting alternative to increase the buffering capacity of the agroecosystem against variations in the weather and the economy. In addition, ICL adds diversity to the entire production system. In this chapter, the authors describe a regional study was conducted in the northern Pampas of Argentina in order to compare soil quality and condition at proximal cropland sites managed under CC and ICL systems under ZT, in representative soils of the region (Typic Argiudoll *vs.* Typic Hapludoll). It is well known that cattle trampling causes damage over the soil physical properties. The natural recovery of the soil physical properties comprises natural and anthropogenic forces, representing the balance of soil amelioration and degradation processes. Natural soil amelioration processes include wetting-drying cycles, swelling-shrinkage cycles, whereas anthropogenic forces include passage of vehicles, tillage, and animal trampling. The soil physical condition naturally recovers to the greatest extent when animals are completely excluded from the plot. In this work, the recovery processes were evaluated in an ICL system, by measuring the dynamics of the soil physical properties over four years. Treatments consisted of closure and grazing plots, in soils with contrasting superficial textures. Introduction of livestock into croplands can lead to shallow soil compaction, which is caused by the collapse of macropores and the deformation of soil structure. Soil bulk density (BD) is usually increased by grazing, as shown in the occurrence of classical, shallow compaction processes. Grazing-induced decreases in BD are less common and, when they do occur, they are not easily explained. They have been ascribed to the intensive transit of animals on saturated soil. Thus, it has addressed the relationship between BD and soil water content (θ) in winter grazing of crop residues in a ZT system. In short, the authors analyzed the effect of cattle trampling at different scales and over time, and particularly considered the decrease in BD after grazing.

Chapter 10 - A mosaic-like spatial pattern, consisting of three main surface cover types (patches) – shrubs, trampling routes, and interspersed open areas – characterizes the landscape of the northern Negev region of Israel. This area has been grazed by flocks of sheep and goats since prehistoric times, and the presence of livestock has created a network of trampling routes which account for over 21% of the surface area of the landscape. The soil characteristics and hydrological responses of these routes differ from those of the remainder of the intershrub area and define a separate microhabitat. Each patch creates specific microenvironmental and functional conditions, which affect the eco-geomorphic system, i.e., relations among soil, water, plants and erosion, therefore, patches may differ from one another in their hydrological characteristics.

The impacts of grazing flocks on the distributions of overland flow and sediment yield, and on soil properties on south- and north-facing hillslopes in a semi-arid area were investigated. Overland flows and sediment yields, and also soil properties, were measured in small runoff plots (0.25–1.0 m^2) representing the three patch types, inside and outside 15-year-old long-term grazing exclosures.

Grazing reduces the spatial heterogeneity of overland flow and sediment yield, increases their total amounts and intensifies the source/sink relationships.

In: Grazing Ecology: Vegetation and Soil Impact
Editor: José Ramón Arévalo
ISBN 978-1-62100-436-3

Chapter 1

GRAZING AS AN ECOLOGICAL FORCE

José Ramón Arévalo*
Department of Ecology, Faculty of Biology, University of La Laguna.
La Laguna (Tenerife), Spain

Goat grazing management is one of the most important management activities worldwide, and it has been a traditional sustainable land use in many areas of the planet (Milchunas *et al.,* 1988; Crawley, 1997), requiring correct grazing management to maintain species composition and high diversity values of the plant communities in grazed lands (Baldock *et al.,* 1994; Olff and Ritchie, 1998). In spite of this traditional and sustainable use of goat grazing, the effects can cause remarkable and significant variation in species composition (Casado *et al.*, 2004; Arévalo *et al.*, 2007).

Grazing effects on species richness and species composition are contradictory when comparing studies (Olff and Ritchie, 1998), and we can find research concluding that herbivores can enhance, have weak effects, or even have a negative impact on plant diversity and richness (Osem *et al.,* 2002; de Bello *et al.*, 2007), as well as different effects on soil nutrients (Bakker *et al.*, 2004; Peco *et al.,* 2006; Fernández-Lugo *et al.,* 2009). The lack of consistent responses among studies has been attributed to factors such as the evolutionary history of grazing, productivity gradients, or grazing intensity (Milchunas and Lauenroth, 1993). The high context-dependence of grazing effects on ecosystems highlights the need to carry out studies that allow us to identify the intrinsic dynamics of the pastures in each region (Perevolotsky and Seligman, 1998).

It is well known that grazing management causes damage to soil physical properties (Fernández et al. *chapter 9*). The natural recovery of soil physical properties comprises natural and anthropogenic forces, representing the balance of soil amelioration and degradation processes. Natural soil amelioration processes include wetting-drying cycles and swelling-shrinkage cycles, whereas anthropogenic forces include passage of vehicles, tillage, and animal trampling. The physical conditions of soil naturally recover to the greatest extent when animals are completely excluded from the plot. However, given the vital importance for the production of livestock and the environmental changes, it is crucial to have a thorough

* Email: jarevalo@ull.es

understanding of the mechanisms that maintain or change the ecosystem in response to the changes of land management. In order to develop sustainable management strategies, it is essential to investigate the processes emerging from pasture practices. Since many rangelands are located in arid or semi-arid areas, water is one of the key variables determining the fate of ecosystems in such regions. Therefore, understanding hydrological processes is critically important (Sugita *et al.,* 2007).

For the Negev desert, in Israel, the impact of trampling routes uses by livestock on soil characteristics is evident. Soil characteristics and hydrological characteristics are different in the routes used by livestock compared to areas without grazing pressure, with trampling increasing the small-scale fragmentation of ecological area constituting a separate microhabitat. In that study (Pariente *chapter 10*), trampling routes played an important role in ecosystem functioning.

The chapters of this book will have different perspectives on the impact of grazing on vegetation, as well as in soil properties, including models that try to clarify these issues. As has been expressed in many reports (Díaz *et al.*,2007), it is necessary to individually asses the importance of grazing at local levels in order to manage specific areas properly and provide the best results to optimize productivity. Grazing management is an old and traditional practice, which is increasingly being abandoned in spite of subsidies for its preservation in the European Community as well as in other countries. Although some studies have shown a negative effect of grazing on endangered plant species (Campbell and Donlan, 2005; Carrete *et al.,* 2009; Gangoso *et al.,* 2006), these results cannot be broadly extrapolated to the effects of traditional grazing systems which exert low pressure on plant communities.

The chapter by Mertia and Santra (*chapter 2*) revises the impact of grazing on vegetation and soil in the Indian Thar desert at different levels They conclude that there is a variety of grazing practices in the area and recommend rotational and deferred rotational grazing systems over continuous grazing, since the first offers higher biomass growth, and provides better opportunities for palatable species, to the grazing animals. Specifically they suggest a grazing base at 70% of carrying capacity, favoring a specific species composition as well as soil conservation.

For some areas, the impact of vertebrate grazers that were introduced during the XIX century on vegetation and soil conservation is clear. In the case of the Australian Alps, the impact of grazers is evident in a reduction of palatable species, and an increase of erosion in wetland area as a consequence of trampling. In areas where no palatable species are present, the impact of grazers is minor and has little effect on ecological processes. However, the reduction of fuel loads due to consumption by grazers can lead to problems in effective control of wildfires (McDougall *et al., chapter 3*) and can be ineffective for fire mitigation.

In the case of Pali Aike National Park, a Patagonian steppe ecosystem, the impact of introduced herbivores (not just goats, but also sheep and ñandús, among others) is considerable for native species, and grazing exclusion experiments have demonstrated that grazing protection increases the success of native species relative to exotic species. Also, native plant communities respond very quickly to grazing exclusion, rapidly recovering the native community at all scales. However, the impact of grazing was not detected in soil nutrient composition, probably due to the low density of grazers. In any case, the pernicious effect of grazing was demonstrated in this area (Domínguez and Pauchard *chapter 5*).

In the Canary Islands, we have a different situation compared to the previous study, because it is a Mediterranean area and, although grazers are introduced, the introduction date was more than 2500 years ago (de Nascimento *et al.* 2009). In this chapter some of the conflicting results on the impact of grazing on Canarian ecosystems are presented, and the authors offer some explanations for these contradictions. First, there are two kinds of areas that have been analyzed to study the effects of grazing: those carried out in currently "pristine" ecosystems, and those done in traditionally managed ecosystems. The first type of area has a high degree of endemism and extant populations of endangered species. Endemic species of the Canarian flora have evolved without herbivore pressures, and may have no adaptations to it (Atkinson, 2001), which would explain their higher sensitivity to goat grazing. Moreover, endangered species, in general, have suffered reductions in their populations due to a combination of anthropogenic factors (removal of woody species, fragmentation of ecosystems, fire, agriculture, and farming), and, being in a very sensitive state, any damage, such as that from an herbivore, can be clearly negative. The second explanation for contradictory findings in the Canary Islands is the broad range of climatic conditions existing on the Archipelago. Conditions vary from arid to humid and so determine the existence of different types of vegetal communities. Both factors, the type of plant community studied (Briske and Noy-Meir, 1989) and the presence of sharp environmental gradients (Milchunas *et al.,* 1988; Osem *et al.,* 2002), greatly impact the response of vegetation to grazing. The third explanation is that the extensive grazing systems, the intensive grazing systems, and the presence of feral goats (differently represented in the reviewed works) can exert varying degrees of pressure on the areas studied (Bermejo, 2003). Based on these collective results, the authors suggest the promotion of goat grazing in traditionally managed ecosystems, applying a strict control of grazing pressure and type of management, as a way of maintaining landscape use, cultural values, and biodiversity under typical grazing pressure (*chapter 4*).

Apart from an important tool to manage species composition and species richness for conservation purposes, grazing can have the important value of controlling invasive species. In the case of the yellow starthistle (*Centaurea solsistialis*), the most widely invasive weed in California, it has been demonstrated that grazing with sheep can be an effective management practice to decrease the abundance of the starthistle, as well as control its expansion. The studies that have been developed and that are shown in the chapter by Álvarez (*chapter 6*) also demonstrate that consumption in specific periods of the year by the sheep have the advantage of the toxicity being null.

Another important part of this book is the modeling of grazing impacts. In the case of studies developed in the French Atlantic coast, the authors (Sabatier *et al., chapter 8*) reveal that it is possible to reconcile conservation and productivity. The best results are obtained when a diversity of management regimes are applied, favoring the best relationship between productivity and conservation. Although they recognize limitations of the model related to the lack of specific information, their first observations in a field study were consistent with the theoretical lines proposed. Modeling has also been used to determine the impact of grazing on soil structure over long time periods and at different intensities in a region of China (Inner Mongolia Grassland Ecosystems Research Station, Northern China). The authors concluded that heavy grazing, deteriorated soil functions, as indicated by decreased total and larger porosity accompanied by decreased saturated hydraulic conductivity and increased bulk density. Among other impacts, it was revealed that intense grazing not only reduced the amount of water available to plants, and thus grassland productivity, but it also possibly increased the risks for wind and water erosion (Zhao *et al., chapter 7*).

In this book we demonstrate the poor extrapolation capacity that grazing studies have, due to site-dependent effects and sensitivity to changes in weather conditions and soil characteristics. So, researchers should be cautious when suggesting management recommendations, such as removing all grazing animals or reducing grazing activities. A highly variable ecological force such as grazing, with important environmental and social implications, should be analyzed deeply before extrapolated results or suggesting management practices.

REFERENCES

Arévalo, J.R., Chinea, E. and Barquín, E. (2007) Pasture management under goat grazing on Canary Islands. *Agriculture, Ecosystems and Environments, 128*, 291-296.

Atkinson, I.A.E. (2001). Introduced mammals and models for restoration. *Biological Conservation, 99*, 81-96.

Bakker, E.S., Olff, H., Boekhoff, M., Gleichman, J.M., and Berendse, F. (2004) Impact of herbivores on nitrogen cycling: contrasting effects of small and large species. *Oecologia, 138*, 91-101.

Baldock, D., Beaufoy, G. and Clark, J. (1994) *The Nature of Farming—Low Intensity Farming Systems in Nine European Countries*. Institute for Environmental Policy, London.

Bermejo, L.A. (2003) *Conservación de los recursos genéticos caprinos en los espacios protegidos de Canarias: Impacto social y ambiental*. (PhD thesis), Universidad de Córdoba, Spain.

Briske, D.D., and Noy-Meir, I. (1998) Plant response to grazing: a comparative evaluation of annual and perennial grasses. In V.P. Papanastasis, and D. Peter (Eds.), *Ecological Basis of Livestock Grazing in Mediterranean Ecosystems* (pp. 13-26). Luxembourg: European Commision, Science, Research and Development.

Campbell, K., and Donlan, C.J. (2005) Feral goat eradications on islands. *Conservation Biology, 19,* 1362-1374.

Carrete, M., Serrano, D., Illera, J.C., López, G., Vögeli, M., Delgado, A., and Tella, J. (2009) Goats, birds, and emergent diseases: apparent and hidden effects of exotic species in an island environment. *Ecological Applications, 19,* 840-853.

Casado, M.A., Castro, I., Ramírez-Sanz, L., Costa-Tenorio, M., de Miguel, J.M. and Pineda, F.D. (2004) Herbaceous plant richness and vegetation cover in Mediterranean grasslands and shrublands. *Plant Ecology, 170,* 83–91.

Crawley MJ (1997) Plant-herbivore dynamics. In: Crawley MJ (Ed) Plant Ecology, Blackwell Science, Oxford, pp 401-474

de Bello, F., Lepš, J. and Sebastià, M.T. (2007) Grazing effects on species-area relationship: variation along a climatic gradient in NE Spain. *Journal of Vegetation Science, 18*, 25-34.

De Nascimento, L., Willis, K.J., Fernández-Palacios, J.M., Criado, C. and Whittaker, R.J. (2009) The long-term ecology of the lost forests of La Laguna, Tenerife (Canary Islands). *Journal of Biogeography*, *36*, 499-514.

Díaz, S., Lavorel, S., McIntyre, S., Falczuk, V., Casanoves, F., Milchunas, D. G., Skarpe, Ch., Sternberg, G.R., Noy-Meir, I., Landsberg, J., Zhang, W., Clark, H., and Campbell,

B. (2007) Plant traits responses to grazing: a global synthesis. *Global Change Biology, 13,* 313-341.

Fernández-Lugo, S., de Nascimento, L., Mellado, M., Bermejo, L.A, and Arévalo, J.R. (2009) Vegetation change and chemical soil composition after four years of goat grazing exclusion in a Canary Islands pasture. *Agriculture, Ecosystems and Environment, 132,* 276-282.

Gangoso, L., Donázar, J., Scholz, S., Palacios, C.J., and Hiraldo, F. (2006) Contradiction in conservation of island ecosystems: Plants, introduced herbivores and avian scavengers in the Canary Islands. *Biodiversity and Conservation, 15,* 2231- 2248.

Milchunas, D.G., Sala, O.E., and Lauenroth, W.K. (1988) A generalized model of the effects of grazing by large herbivores on grasslands community structure. *The American Naturalist, 132.* 87-106.

Milchunas, D.G., and Lauenroth, W.K. (1993) Quantitative effects of grazing on vegetation and soil over a global range of environments. *Ecological Monographs, 63,* 327-366.

Olff, H., Ritchie, M.E. (1998) Effects of herbivores on grassland plant diversity. Trends in Ecology and Evolution 13, 261-265.

Osem, Y., Perevolotsky, A., and Kigel, J. (2002) Grazing effect on diversity of annual plant communities in a semi-arid rangeland: interactions with small-scale spatial and temporal variation in primary productivity. *Journal of Ecology, 90,* 936-946.

Peco, B., Sánchez, A.M., and Azcárate, F.M. (2006) Abandonment in grazing systems: Consequences for vegetation and soil. *Agriculture, Ecosystems and Environment, 113,* 284-294.

Perevolotsky, A., Seligman, N.G., 1998. Role of Grazing in Mediterranean Rangeland Ecosystems. Inversion of a paradigm. Bioscience 48, 1007-1017.

Sugita, M., Asanuma, J., Tsujimura, M., Mariko, S., Lu, M.-J., Kimura, F., Azzaya, D., and Adyasuren, T. (2007) An overview of the rangelands atmosphere–hydrosphere–biosphere interaction study experiment in northeastern Asia (RAISE). *J. Hydrol. 333*, 3–20.

In: Grazing Ecology: Vegetation and Soil Impact
Editor: José Ramón Arévalo
ISBN 978-1-62100-436-3

Chapter 2

GRAZING PRACTICES IN THE RANGELANDS OF THE INDIAN THAR DESERT AND ITS IMPACT ON ECOSYSTEM AND ENVIRONMENT

***R. S. Mertia*[1] *and P. Santra*[2*]**
[1]Principal Scientist and Head, Central Arid Zone Research Institute, RRS, Jaisalmer, India
[2]Senior Scientist, Central Arid Zone Research Institute, Jodhpur, India

ABSTRACT

In the Indian Thar desert, different type of grazing practices is followed since long. Scientist from Central Arid Research Institute (CAZRI), Jodhpur has conducted several experiments on grazing and its impact on livestock, ecosystem and environment. It has been found that the ever increasing trend in the livestock and human population has put heavy grazing pressure and continuous decrease in the area of land available for grazing. Due to these factors, around 60% of grazing lands are in poor condition, 14% in fair, 13% in good and 2–3% in excellent state. Low and erratic rains and frequent drought restricts the vegetative and self seeding period of natural vegetation. The different conventional grazing experiment has provided useful information about the impact of grazing on rangelands of the Indian Thar desert. It has been observed that continuous grazing led into reduction in the density, basal cover and forage production of many palatable perennial species and increase of unpalatable species. The rotational and deferred rotational grazing systems are superior over the continuous controlled grazing. It has been observed these systems provide rest to the perennial species, which can assume their growth and vigor and produces high biomass and also these systems provide better opportunities to the grazing animals for utilizing animals and those species which become less palatable after maturity.

While, taking decisions on stocking rate of animals in rangeland it also should be kept in mind that the rangeland should not be overgrazed. Because, experimental studies have shown that overgrazing may lead to 3 to 5 times higher soil loss than controlled grazing due to wind erosion. Moreover, these eroded soils pose multifaceted problems in

---

* Corresponding author Email: priyabrata.iitkgp@gmail.com

environment e.g. heavy atmospheric load of health hazardous fine dusts, loss of soil carbon and nutrients etc.

Therefore, grazing resources of the Indian Thar desert are to be managed in a sustainable manner specifically under the climate change scenarios and under increasing trend of grazing pressure in rangelands. Here, we have reviewed the existing grazing resources of the Indian Thar desert, livestock-pasture relationship under different grazing systems of the region, impact of different grazing systems on ecosystem and environment. This will help to formulate suitable rangeland policies in the Indian Thar desert with an aim to control the grazing pressure and to obtain sustained primary and secondary productivity.

INTRODUCTION

The Indian Thar desert may be divided into two climatic zones based on rainfall pattern: arid zone having annual rainfall <250 mm and semi-arid zone with annual rainfall ranging between 250 and 500 mm. Grazing based animal husbandry is the predominant occupation of peoples of arid region of the Indian Thar desert. However, livestock based economy has a subsidiary role in semi-arid regions, where dry farming is the major occupation. This is quite evident from the land coverage pattern of grazing lands in these two regions. In the semi-arid zone, the rangelands occupy 15-20% lands whereas it is around 45% in arid region. In extreme arid tracts of the Indian Thar desert, grazing lands even occupy 85-90% of the total area. Moreover, grazing lands in semi-arid zone are mostly small in sizes and located on marginal or degraded lands, which are earmarked as village community grazing lands.

The increase in livestock density from 13.4 km^{-2} in 1956 to 27.91 km^{-2} in 2001 coupled with many fold increase in human population has caused serious reduction in grazing lands viz. pastures from 8.89 million ha in 1956 to 6.84 million ha in 2001 and fallow land from 10.2 million ha in 1972 to 5.76 million ha in 2001 and thus has tremendously increased the grazing pressure on existing resources. In desert regions, the accessible tracts are being overgrazed continuously and as a result the range condition is poor with a carrying capacity of only 0.2 to 0.5 adult cattle unit (ACU) ha^{-1} (Mertia, 1992a). In spite of poor range condition, the existing grazing pressure in these rangelands ranges from 1 to 4 ACU ha^{-1}. Therefore, the vegetation in the Indian Thar desert is commonly shaped by nomadic and sedentary form of pastoral use. The palatability of perennial and annual components of the rangeland biomass to the livestock also varied widely within this desert area. As such, overgrazing enhances the competition among livestock species for grazing of more palatable plant species in the rangeland (Mertia, 1996). Therefore, the existing situation of rangelands at both arid and semiarid parts of the Indian Thar desert portraits the past misuse and present neglect on management of these important rangelands (Mertia, 1992a).

Moreover, due to adverse physical and climatic conditions, the production of grasslands in the desert area is lower than that in temperate or tropical areas. Less production results in scarce vegetation coverage and poor spatial structure and floral composition. Under these circumstances, the irreversible changes in plant species composition or soil structure should occur if the grazing by animals is higher than the annual regeneration capacity of vegetation (Crawley, 1983). Even under the conditions where no irreversible changes of vegetation coverage occur, grazing by animals changes plant community structures and dynamics through the different effects of grazing behavior of different animal species. Selective feeding influences dynamics of each component heterogeneously. The number and viability of certain plant species is reduced by feeding and another unpalatable or grazing-tolerant species is increased.

Rangelands of the Indian Thar desert also protect the loose surface soil to erode by wind by providing surface cover. However, heavy grazing pressure on the rangelands causes the minute soil particles to erode at higher rates and thus cause several environmental problems e.g. contributing to aerosol load, increase in hazardous particulate matter in atmosphere, loss of soil carbon etc. Therefore, grazing practices in the Indian Thar desert needs to be managed properly to protect the desert ecosystem and also to safeguard the future desert environment.

THE INDIAN THAR DESERT

Location

The Thar desert forms the eastern limit of a vast arid tract of arid region that encompasses the desert of Sahara, the Arabia, Iran, and Beluchistan in Africa and Asia. The eastern limit of the Thar is roughly coterminus with the Aravalli hill ranges-one of the oldest surviving in the world, while its western margin is along the fertile plains of the Indus in Pakistan. In the south it has a sharp natural boundary with the world's largest saline waste-the Great Rann of Kachh, while in the north the riparian sub Himalayan plains define its boundary (Figure 1).

Figure 1. Location of the Indian Thar desert.

It is situated between the latitudes of 24°30′ N and the longitudes of 69°30′ N. Within the territory of India the Thar forms a part of the country's sandy hot arid zone, spread over 0.32

million sq. km area. Western Rajasthan alone constitutes (61%), while in the state of Gujarata it covers 20% area and in the state of Punjab and Haryana 9% area. The southern state of Maharastra, Andhra Pradesh and Karnataka together contribute the rest 10% area.

Climate

The climate is arid, where precipitation over a greater part of the year is less than the potential evapotranspiration and meets less than one third of the annual water need. A significant feature of the distribution of the rainfall is received during the south west monsoon period, June to September. The contribution of this seasonal rainfall to annual rainfall is quite high (91-96%) over whole of Gujarata and south and central parts of western Rajasthan (Ramakrishna *et al.,* 1992). Another interesting feature of the rainfall is the higher coefficient of variability in the annual rainfall, which often exceeds 50% in north west Indian arid zone and is higher than 70% in the extreme region of western Rajasthan where the annual rainfall is as low as 180 mm. Such a poor rainfall performance is due to low level divergence and associated subsidence of air up to 0.8 km, as also marked divergence at a height of 2 to 6 km over the region. Dust storms are frequent during hot weather season, especially in the month of May and June. The average wind speed during June is more than 25 km hr^{-1} in the western part of arid regions. Annual rainfall along with minimum and maximum temperature during last 30 years at Jaisalmer and Chandan located at western part of the Indian Thar desert is presented in Figure 2.

The mean diurnal variation of temperature is 14°C to 16°C. The mean maximum temperature during the season varies from 40°C to 42°C but in some years mercury has touched 50–52°C. During the cold season minimum temperature varies from 3°C to 10°C. Frost occur at all stations when temperature touches –2°C to –4°C, but its incidence varies from 5 to 20 days and is associated with the western disturbances bringing in cold.

Soil

Arid zone soils of India extend through an area of about 200,000 sq. km and these occur predominantly in the region of Thar desert in India. Ray Choudhari et al. (1963) described broadly the soils of arid zone as “Desert soils” in the west and the “Grey brown soils” in the south east. Roy et al. (1978) divided the soils into “desert soils” and “sand dunes” because unlike “Desert soils”, “Dune soils” did not show any lime concretionary zone in the profile and they occurred in association with “Red desert soils” also. Soils of arid regions of this region have further been put into major six groups by Dhir (1977).

The surveying efforts of several soil profiles in India by National Bureau of Soil Survey and Land Use Planning (NBSS and LUP) had resulted in the soil survey reports for different states. Along these series of studies, soil profiles from the Indian Thar desert were also explored and reported in the form of soil series database of Rajasthan and Gujarata (Shyampura *et al.,* 2002, Sharma *et al.,* 2006). It was found from the database that soils from the Indian Thar desert belong to mainly three soil orders: aridisol (67%), entisol (24%), and inceptisol (9%). Aridisols were equally distributed within the region.

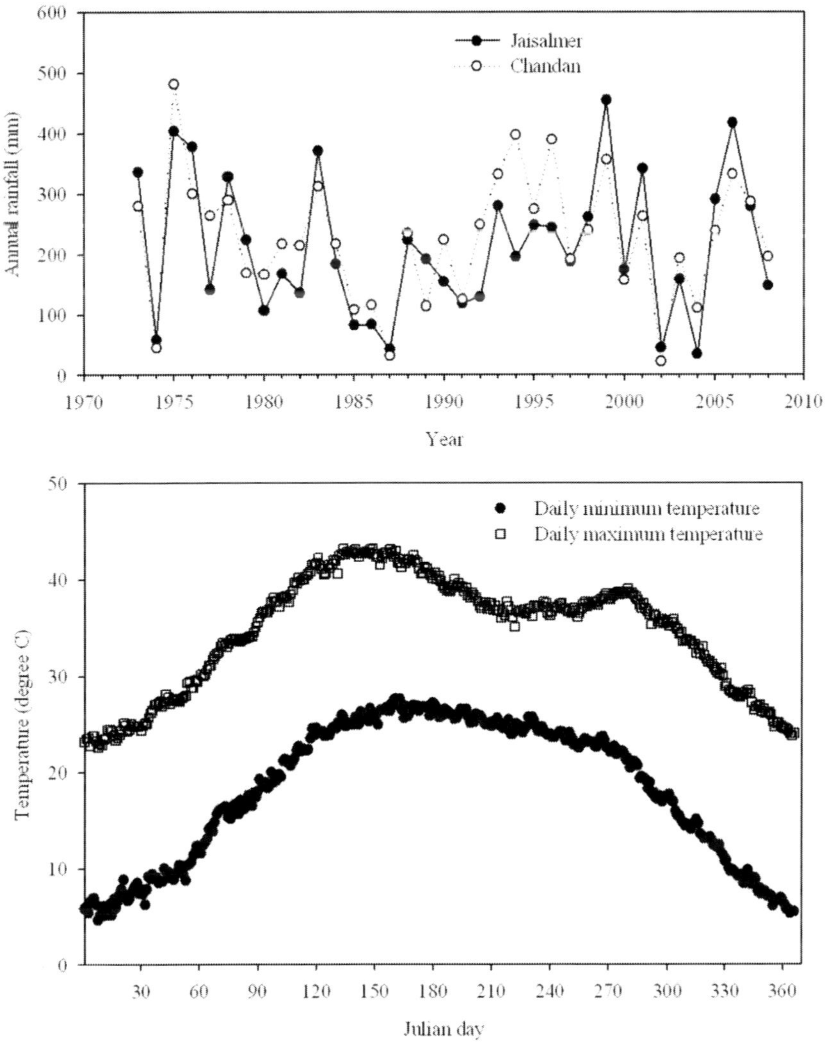

Figure 2. Long term annual rainfall and daily minimum and maximum temperature in the Indian Thar desert (Data source: Daily meteorological data of Jaisalmer and Chandan station of CAZRI, Jodhpur from 1981 to 2009).

Entisols were dominantly located at the western and northern plains of the desert, whereas inceptisols were dominantly located at the south eastern part of the desert. Characteristically, accumulation of $CaCO_3$ at shallow depth was observed for most of the soil profiles under aridisol. Soils under entisols were concentrated at those places where wind erosion/deposition processes were very active and dunes and steep slopes were most common landscape features. Soils under inceptisols were mostly occurred in southern coastal part of the desert, where aridity was not so severe. An overview of soil properties from the Indian Thar desert are given in Table 1. Sand content of the soil samples were high with a mean value of 63.87%. The average silt and clay content of soil samples were 14.22% and 22.92%, respectively. Soil textural class was sandy for most of the cases. Few soil samples with clay and sandy clay soil texture were also observed, which belonged to soils from southern coastal region of the desert. Organic carbon content of soils was low with an average value of 2.55 g kg^{-1} of soil. The average soil water content at 1/3 bar ($\theta_{1/3bar}$) was 17.49% on volumetric basis and varied widely from a very low value of 2.50% to as high as 47.70%. The average soil water content at 15 bar (θ_{15bar}) was 7.04%. Wide variation in θ_{15bar} was also observed with a minimum value of 1.20% and a maximum value of 19.40%.

Table 1. Descriptive statistics of soil properties from the Indian Thar desert

Soil properties	Mean	Standard deviation	Range
Sand (%)	63.87	22.64	3.50 – 97.00
Silt (%)	14.22	11.15	0.40 – 67.90
Clay (%)	22.92	14.37	1.00 – 68.40
OC (%)	2.55	1.94	0.1 – 11.00
$^{\S}\theta_{1/3bar}$ (%, m^3 m^{-3})	17.49	9.39	2.50 – 47.70
$^{\S}\theta_{15bar}$ (%, m^3 m^{-3})	7.04	4.17	1.20 – 19.40

$^{\S}\theta_{1/3bar}$ and θ_{15bar} represents the volumetric water content (%, m^3 m^{-3}) of soil at 1/3 bar and 15 bar, respectively.

Flora

Plant life in the Thar desert have evolved through many evolutionary phases. The occurrence of fossils of *Callophyllum*, *Garcinia*, and *Mesua* (Lakhanpal and Bose, 1951) and *Cocos* (Kaul, 1951) in Tertiary (Eocene) sediments indicated humid environment of tropical rainforest. The changes in the plant life from cool temperature of the permo-carboniferous to tropical in the upper cretaceous suggest gradual northward movement of the Indian plate across the equator (Lukose, 1984).

The sequence of events leading to the change from above to the xerophytic form of the holocene becomes difficult to trace as no paleobotanical records belonging to 50 million years that elapsed between those two epochs are available. It is, therefore, believed that "the present day flora was well established at the beginning of holocene (10,000 years ago) and that subsequent changes in the post-glacial vegetation were more ecological than floristics" (Lakhanpal and Singh, 1975).

There has been a high intervention of biotic activities, which has influenced true equilibrium of flora and climate. The present, physiognomy is, therefore more xeric than the climate warrants (Shankar, 1994).

The flora of the Indian desert is constituted by the following life forms viz; (Th/40.1%, Ph/22.1%, Ch/18.9%, H/15.5%, Cr/3.4%) (Das and Sarup, 1951). The percentage of the therophytes further increases to 49% in the western hyper-desertic region (Mertia and Bhandari, 1991). The most characteristic life forms of the Indian desert are chamaephytes (18.9%), which are the leading species, of most association. The main species are *Fagonia cretica*, *Tephrosia uniflora*, *Rhynchosia chimperii R. pulverulento*, *Alysicarpus longifolius*, *A. moniliter* etc.

Grazing Resources and Grazing Lands in the Indian Thar Desert

Major portion of the Thar desert (61%) lies in the western Rajasthan and a little (20%) in the western part of Gujarata state. Grazing resources of arid and hyper-arid regions have been well documented by the scientists of Central Arid Zone Research Institute (CAZRI), Jodhpur by conducting research at 52 range management and soil conservation centres over last 50 years. The rangelands in this deserts includes grasslands, forest grazing areas and wastelands of various categories (Figure 3) and these generally support different species of perennial grasses with scattered thorny trees, bushes and shrubs and some of them have high forage value. The area under grazing land have declined by 23.78% from 1951 to 1971-71 and is estimated to have further declined during last 30 years with a increasing pressure on grazing resources. Mertia (1992b) have clearly demonstrated on the basis of 20 years of research on rangeland improvement and utilization that grazing conducted on basis of carrying capacity can increase the productivity from poor to very good over a period of 15 years even in the hyper arid regions of Thar desert.

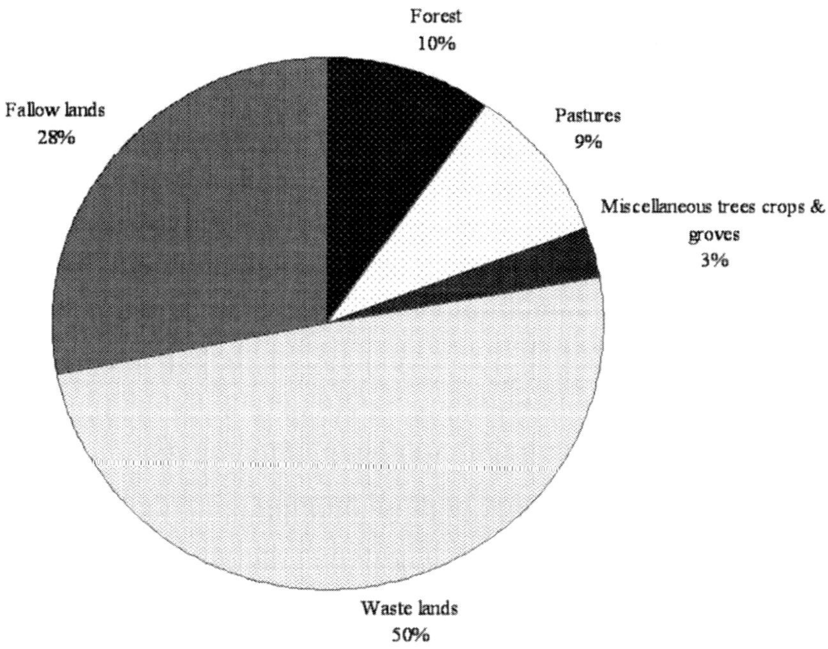

Figure 3. Grazing resources in the Indian Thar desert.

The first country wide report of the grazing lands was published by the ICAR in the form a book entitled "Grass covers of India", authored by Dabadghao and Shankarnarayan (1973). Five major grass covers, their aerial extent and herbage yield had been described by them. The tropical *Sehima-Dichanthium* cover is the largest followed by sub-tropical *Dicanthium-Cenchrus-Lasiurus*, *Pharagmites-Sacharaum-Imperata* and *Themeda-Arundinella* and temperate and alpine meadows in the Himalays and Nilgiris. The only climatic climax of grassland vegetation in India is the temperate alpine meadow, which are largely dominated by festuwid grasses. The tropical and subtropical grasslands are derived tertiary serial communities (Whyte, 1964) where the succession processes are arrested due to continued biotic interferences viz, grazing and or burning (Champion, 1936). The flora of grazing lands is dominated by therophytes (Mertia, 1976; Mertia and Bhandari, 1992). The preponderance of therophytes results from strong periodicity in tropical monsonic climate. The trees and shrubs are usually associated with grasslands.

The growth of vegetation in the Indian desert starts with the onset of monsson in the end of June or beginning of July, each year. The top growth is attained in September or October and then it declines upon maturity as most of ephemerals have very short life cycles. The period of active growth depends on species composition.

The above ground production in the semi-arid regions in Jodhpur (350 mm yr^{-1}) has been estimated at 1.2–2 t ha^{-1} yr^{-1} whereas in the driest region, Jaisalmer (180 mm yr^{-1}) the dry herbage yield of 2-2.5 t ha^{-1} has been recorded (Mertia, 1992a).

Common property lands (CPLs) consisting mainly of barren lands and long fallow have been declining at the rate of 2.24 and 1.93% per annum in the regions of the Thar desert. The present average forage production has been estimated 0.295 t ha^{-1}. The house hold figures revealed that stocking rates at farm level were larger by 24 to 32% compared to optimum on different sizes of farm (Anantha Ram, 1993).

The main reasons of degradadtion of CPLs include (i) permanent grazing of animals by landless and small farmers (ii) overgrazing and heavy trampling by animals every day while coming for drinking water and shelter for some period. In fact no grass species grows to the stage of maturity in these lands due to very heavy stocking, which has converted these lands into highly degraded conditions. The most palatable *C. setigerus*, *Dactyloctenium sindicum*, *Cenchrus biflorus* and *Eleusine compressa* are grazed to ground level and are even uprooted and have disappeared completely. The common unpalatable species viz. *Tephrosia purpuria*, *Aerva persica* and *Leptadaenia pyrotechnica* are found. Trees viz. *Ziziphus rotundifolia*, *Prosopis cineraria*, *Maytenus emarginatus*, *Salvadora oleoides* and *Capparis decidua* are commonly growing on such sites.

DYNAMICS OF RANGELAND VEGETATION

The grassland in semi-arid region has richer flora than in the arid region, total number of species observed is 64 species in our experimental field in semi-arid site (Pali), and 34 species in arid site (Chandan). In the semi-arid site, non-leguminous forbs are prevalent more than half (34 species), this value is more than double as that of arid site. Grass species were more dominant (34%) in Chandan than Pali (25%). In view of species number, the percentage of legumes, trees and shrubs, and annual/perennial ratio is almost same in both semi-arid and arid regions.

Seasonal Change in Coverage

The total coverage of grassland changed significantly in the course of the season in both arid and semi-arid region (Narita *et al.*, 1996) (Figure 4). Especially in the arid region

(Chandan), the change was conspicuous. Comparing the change in control plot, there is distinct difference between two situations. In arid region (Chandan), the coverage increased rapidly from 5% to 35% from August to September then decreased quickly. On the contrary, in the semi-arid region (Pali), there was no such tendency; the total coverage was steady around 15%. The reason for this difference was mainly owing to the coverage of legumes. In Chandan, coverage of legumes rose from 0% in January to 29% in August; it contributed 89% to total coverage. In Pali, legumes grew to only a few percentages. The coverage of grasses in Pali was more fluctuated than in Chandan, decreasing from January to September. The coverage of non-leguminous forb was quite small and constant in Chandan, but that values in Pali gradually increased from January to September, and then decreased in October.

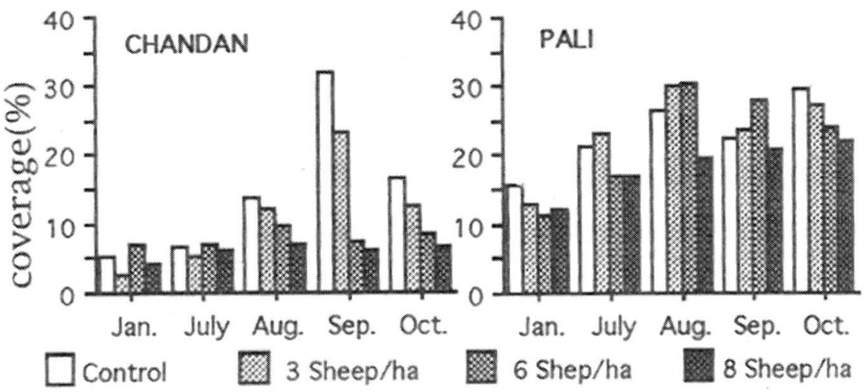

Figure 4. The seasonal change of total grassland coverage and effect of grazing in arid (Chandan) and semi-arid region (Pali) of the Indian That desert (Source: Narita *et al.*, 1996).

In respect of the life form, there were same trends in differences. In Chandan, perennials were quite stable, and change of annuals was conspicuous. On the other hands, the changes of both annuals and perennials were not outstanding.

Seasonal Change in Diversity

Diversity in rangelands of the Indian Thar desert was reported by Narita et al. (1996). Diversity index was calculated as Shannon-Wiener function (Shannon and Weaver, 1949). The diversity index increased quickly at the beginning of the monsoon season in both arid and semi-arid region. The values were higher in semi-arid region (Pali) than in arid region (Chandan) throughout observation. During monsoon season, the value did not vary in semi-arid region, but in arid region, it had moderate peak in August.

GRAZING SYSTEMS IN THE INDIAN THAR DESERT

Different types of grazing systems practiced in the Indian Thar desert are described in Sharma (1987). All these grazing systems are managed in such a way so that it balances between the needs of the plants and animals in rangeland. A brief description of different grazing systems is mentioned below:

1. *Continuous grazing*: This is an extensive system where stock remains on the same pasture land for prolonged periods. Stocking rate is generally low in this type of grazing system. Selective grazing of animals is inevitable in this system. Therefore, in the dry season selective grazing together with over-grazing may cause the deterioration of the pasture in long run.
2. *Rotational grazing*: In this system grazing area is divided into a number of paddocks (usually at least six) and the herd is moved systematically from one to another paddock in rotation. High stocking rates per paddock may be used. Each paddock is grazed for a period of one to seven days or in some conditions it may extends up to 15 days, depending on stocking rate and herbage growth rate. By the time the last paddock has been grazed, the first should be ready for another grazing. Rest period of two to four weeks should be aimed at for sown pasture.
3. *Deferred grazing*: This is the practice of conserving standing hay, whereby certain paddocks are left ungrazed, for use during the dry season. When deferment is combined with rotational grazing, it is called deferred rotational grazing. In this case deferred block or paddock is changed in rotation every year. Deferred rotational grazing carried out in different range management areas revealed no significant differences in forage yield as well as growth rate of heifers in continuous control grazing and during different rotational periods. However, the growth rate of the yearling heifers was significantly affected by the seasons irrespective of the year and the treatments of grazing. The growth rate of heifers ranged between 48.4 to 88.0 kg animal^{-1} in a year under this grazing management system (Verma and Ahuja, 1978).

STOCKING RATE IN GRAZING SYSTEMS

Rangelands are essentially to be used for optimized livestock production. As such, a proper balance between the number of livestock and productivity of the range need to be maintained by continuous and careful observations on the morphological and physiological characters of plants growing on the range. Proper stocking rate on the range is the first range conservation practice which should be adopted.

'Excellent', 'Good', 'Fair', and 'Very poor' condition rangeland (having approximate productivity of 20, 15, 10, 7.5 and 5.0 q ha^{-1}, respectively) can safely provide year long grazing to 25-30, 20, 17, 13 and 1-6 adult cattle units per 100 ha blocks, respectively during the normal years (Bhimaya and Ahuja, 1969). Under abnormal years, grazing stress has to be increased or decreased depending on the availability of forage on the rangelands.

Experimental results have revealed that stocking rate of 2.4 ha heifer^{-1} was capable of giving 237-270 gm day^{-1} heifer^{-1} body weight gain in case of *Cenchrus* and *Lasiurus* rangelands in the arid zone having less than 300 mm rainfall. In areas getting higher rainfall (between 300-600 mm), stocking rate of 1-3 ha heifer^{-1} was capable of providing almost similar animal body weight gain as reported in arid areas. Similar stocking rate of ramblambs per ha are capable of giving 30-40 gm day^{-1} rangeland^{-1} in arid zone of western Rajasthan (Paroda, 1978).

Studies conducted on different stocking rates without supplemental feeding on different types of rangelands revealed that growth rate per animals remains high when the animals grazed on the carrying capacity basis (2.40 ha heifer^{-1}). Although by increasing the stocking rate viz. 12.20 and 0.60 ha heifer^{-1} the total livestock production goes considerably high but the grass component, particularly the animal species got eliminated through effective utilization. As a result of which, it was considered essential to provide concentrates to the animals during lean periods (December–June) so as to meet out their digestible crude protein (DCP) and total digestible requirements.

IMPACT OF GRAZING

Impact on Livestock under Different Grazing Managements

Several studies were conducted by the range management section of CAZRI on different systems of grazing management in the rangelands representing varying agro-climatic conditions of western Rajasthan and the results were reported in Paroda *et al.,* (1980). Here, highlights from the major results are mentioned below:

Under different systems of deferment, maximum gain in body weight of adult cows (average body weight 270 g) was observed in continuous controlled grazing (based on carrying capacity) system irrespective of different type of rangelands. However adult cow exhibited gains in body weight by 25.8, 35.8 and 56.0 kg on an average in 'Poor', 'Fair', and 'Good' condition rangeland, respectively. Deferment of grazing for eight fortnights from growing period of vegetation resulted in decrease in body weight by about 30 kg per cow. Similarly, adult sheep exhibited body weight gain of the order 9.2 and 7.1 kg per sheep in 'Good' and 'fair' class rangelands, respectively under continuous controlled grazing system. Deferment of grazing on rangelands for 16 weeks from July gave less body weight gain.

Growth of yearling heifers under system of monthly rotational grazing based on carrying capacity on an average gave monthly growth of 7.4 kg heifer^{-1} in *C. setigerus* cover, 6.6 kg heifer^{-1} in *Cenchrus-Sporobolus* cover, 7.9 kg heifer^{-1} in *Cenchrus ciliaris* cover and 4.5 kg heifer^{-1} in *L. sindicus* cover. From July to October, growth rate of animals remained highest on the range and it was of the order of 11.2 to 13.7 kg heifer^{-1} without detrimental effect on the existing range conditions. Growth rate remained highest during July–December. In areas where *Cenchrus biflorus* and *Aristida* dominated, the growth declined mainly during September-October as a result of discomfort due to piercing of awns and burs in the body and mouth parts of grazing animals.

Considering different systems of grazing of both the heifers and sheep, it appeared that the continuous controlled grazing, based on carrying capacity, was equally effective in the rangelands when compared with deferred rotational grazing. However, it has been observed that in dry land ranges, due to continuous grazing for number of years, high yielding and palatable species showed a declining trend. Hence, distinct advantage of deferred rotational grazing for better establishment of perennial grasses through natural self seeding is obvious, especially in areas where reseeding has not been practiced.

Studies on mixed grazing with cattle and sheep on *Lasiurus-Cymbopogon-Aristida* rangeland in rainfall zone below 250 mm was conducted wherein heifers and lambs grazed separately and when both the animals grazed together. Results revealed that the growth of animals per unit area remained to be the highest when heifers grazed alone followed by mixed grazing with heifer and sheep and the least when sheep grazed alone.

Seasonal grazing with yearling heifers on rangelands with rainfall below 250 mm revealed that growth per animals during monsoon (August–October), winter (November–January), Spring (Feb–April) and summer (May–July) was 27.6, 14.0, 10.1 and 4.3 kg per animal respectively giving a total growth of 47.4 kg per animal. Losses in body weight of animals during summer were done due to quality of the forage, high temperature and hot wind speed etc.

Studies on comparative performance of goats and sheep on sown pasture of *Cenchrus* species infested with *Zizyphus numularia* and *Mimosa hanata* bushes revealed 292 percent increase in the body weight gains in buck (male goat) as compared to ramblambs (male sheep) within a period of one year under light intensities of grazing (3 animals ha^{-1}). The increase in gain of buck over ramblamb was 178 and 75 per cent under medium (4 animals ha^{-1}) and heavy (6 animals ha^{-1}) intensities of grazing, respectively.

Impact on Ecosystem

Effect of Grazing on Coverage and Diversity

The seasonal changes in coverage under different grazing pressure were studied by scientists of CAZRI at arid (Chandan) and semi-arid region (Pali) of the Indian Thar desert (Narita *et al.,* 1996). The effect of grazing on coverage of grassland community was found significant in both arid and semi arid region of the Indian Thar desert. In arid region grassland, the effect was more distinct, from August to October. Particularly, in the month of September, the coverage in grassland with grazing pressure of 8 sheep ha^{-1} was less than one sixth of no grazed grassland. While the coverage in no grazed grasslands was 32% in September, the same was only 6% in grasslands with grazing pressure of 8 sheep ha^{-1}. The change in coverage due to grazing pressure is mainly contributed by legumes and annuals. The coverage of grasses does not differ with different grazing pressures. In the semi-arid region, although there was a significant effect of grazing on grassland community, there was no clear trend in the effect of grazing on total coverage. In both arid and semi-arid regions of the Indian Thar desert, grazing did not affect the diversity of grassland community.

Effect on Forage Production

The forage production is significantly affected by different intensities of grazing. Medium and light intensities do not affect the forage production adversely but the heavy or severe grazing affected the forage production of perennial grasses. In a study from arid rangelands, it was reported that at the initial stage of heavy intensity grazing, 86.5% of pasture grasses are utilized (Sharma, 1987). In another study in semi-arid rangeland indicated that heavy intensities of grazing led to 86.59% utilization of edible species. It was also reported that forage yield generally increased with decrease in grazing intensity which could be attributed to higher growth rates observed in medium and light grazing. Fundamentally, favourable development of shoot apex takes place in case of medium and light grazing, whereas the shoot apex is damaged in case of heavy grazing.

Effect on Grass form of Vegetation

The impact of overgrazing is more pronounced on the perennial grasses and trees/ shrubs having suitable height for browse by camels and goats. The dominant grasses *Lasiurus sindicus*, *Panicum turgidum*, *P. antidotale*, *Cenchrus ciliaris*, and *C. setigerus*, which grow erect ('bakket' shape) and extensively in light intensity of grazing turns into 'saucer' shape under heavy grazing pressure. Similarly, browsing combined with lopping and felling in the early phases of growth of dominant species like *Ziziphus mauritiana*, *Prosopis cineraria*, *Capparis decidua*, and *Salvadora oleoides* are turned into bush form, which otherwise are small size trees. This is mainly due to heavy browse by camels and goats, which never allow them to attain their natural form. Pandey (1994) observed that it is the intensity of grazing, which is the decisive factor in shaping the physiognomy and structure of the grassland vegetation.

Effect of Grazing on Population Dynamics

Lasiurus sindicus and *Cenchrus setigerus*, which were the dominant perennial grass species in arid and semi-arid regions, respectively showed different response to grazing. Though *L. sindicus* did not show clear relationship between coverage change and grazing intensity, whereas, *C. setigerus* decreased with grazing gradient. In semi-arid region, two perennial grasses, *C. setigerus* and *Eremopogon foveolatus* has same trend, they had rapid drop during hot summer (from January to July) and marked increase at the end of monsoon season. Two annual legumes, *Indigofera cordifolia, I. linifolia* suffered from most severe grazing in the arid region of the Indian Thar desert. However, in semi-arid region, *I. cordifolia* showed opposite response to grazing. The coverage of *Lepidagathis trinevis* showed clear negative response to the grazing; it decreased gradually with grazing increased. An annual non-leguminous forb, *Tribulus terretris* showed unique response to grazing. In arid region this species increased with increase of grazing intensity, on the contrary, it decreased with increase of grazing in semi-arid region.

Effect on Phenology and Palatability

The impact of different grazing intensities by sheep on the phenology of *L. sindicus* pasture and its relationship with palatability was studied by Kumar et al. (1999). The monsoon rains of July quickly stimulated dormant plants into vegetative growth followed by flowering and seed set. With increasing grazing pressure, *L. sindicus*, the dominat perennial grass with 80% cover showed earlier flowering, fruiting and seed set. Telescoping of the vegetative phase seemed to be a mechanism for evading grazing pressure in this perennial grass. However, the annual species responded otherwise. In the annual species *Aristida funiculate*, *Indigofera cordifolia*, *Indigofera linifolia* and *Corchorus tridense* increasing intensity of grazing prolonged the vegetative phase and thus delayed the completion of phenological cycle. These phonological shifts could be easily related to the palatability trends. *L. sindicus* was 100% palatable just before the onset of monsoon and even after the first few showers. As the annuals came after rains, they were eaten in preference to the perennial *L. sindicus*. In fact, palatability of *L. sindicus* declined to almost nil. While the palatability of most annuals started declining after peak in August, *L. sindicus* showed increasing trend, until it became the dominant palatable grass by November. In the second year of grazing these trends become sharper. It can be surmised that if this trend continues, slowly but surely palatable annual species will be the first to disappear with overgrazing followed by perennials; finally the unpalatable species will be left.

Impact on Environment

Grazing Pressure and Wind Eroded Soil Loss

To study the effect of grazing pressure in rangelands on soil loss from field experiments were carried out at two major land use systems of the Indian Thar Desert (Mertia *et al.*, 2010). Site I was situated at the research farm of the Central Arid Zone Research Institute located at Jaisalmer (26°55′N and 70°57′E). Site II was situated at the research farm of the Central Arid Zone Research Institute located at Chandan (27°01′N and 71°01′E), which was 45 km in North-East direction from the first site. Both sites are situated in the Jaisalmer district of Western Rajasthan, which falls under extremely arid regions of the Indian Thar Desert. Both rangeland sites are dominantly covered with *Lasiurus sindicus* (sewan) grass, which contributes 80–85% of the total biomass production of the rangeland. These two rangeland sites were imposed to different levels of grazing pressure. At site I, grazing was allowed throughout the year with an average stocking rate of 3.2 ha^{-1} (Mertia *et al.*, 2006), whereas at site II, grazing was allowed for 3 months with average stocking rate of 2 ACU ha^{-1} (Mertia *et al.*, 2006). At the controlled grazing rangeland site, grazing was allowed up to a minimum tussock height of 15 cm, whereas in the overgrazed rangeland site, animals grazed the sewan grass up to surface level and uprooted the underground biomass in severe cases. The excessive trampling of grazing animals in the overgrazed site resulted in more loosening of the surface soils. The average density of sewan grass at the overgrazed site was only 5700 tussocks ha^{-1} in comparison to the 8300 tussocks ha^{-1} at the controlled grazing site (Mertia *et al.*, 2006). The average basal cover area by the tussock of sewan grass at the overgrazed site was 0.15–0.20 m^2, whereas at the controlled grazing rangeland site, it was 0.45–0.50 m^2.

Aeolian mass transport (kg m^{-1}) in suspension mode was computed at each site according to power decay mass–height profile model. The time average mass transport rates (kg m^{-1} day^{-1}) during the last week of April to the first week of August, 2004 at each site are depicted in Figure 5. The mass transport rate was almost three times higher at the overgrazed site than at the controlled grazing site during mid of June to mid of July. Notably, the only difference between these two sites was the level of grazing pressure, which resulted in huge difference in mass transport rates. This clearly indicates that simple control of grazing in arid rangelands of the Indian Thar Desert may significantly reduce the active land degradation process due to wind erosion.

Overall, the total amount of soil loss during summer (May–July) of 2004 was 827 kg ha^{-1} at the overgrazed site and 240 kg ha^{-1} at the controlled grazing site (Mertia *et al.*, 2010). The extrapolation of mass–height profile also revealed that on an average 60% of total soil loss occurred in suspension mode (0.3–2 m height) at the overgrazed site, whereas it was 65% at the controlled grazing site. More vegetation coverage at the controlled grazing site comparatively reduced the Aeolian mass flow at near surface layers. However, the mass flow in creep and saltation mode showed an increasing trend with severity of wind erosion and reached to a maximum level of contribution of 58% and 54% to total soil loss at the overgrazed site and the controlled grazing site, respectively.

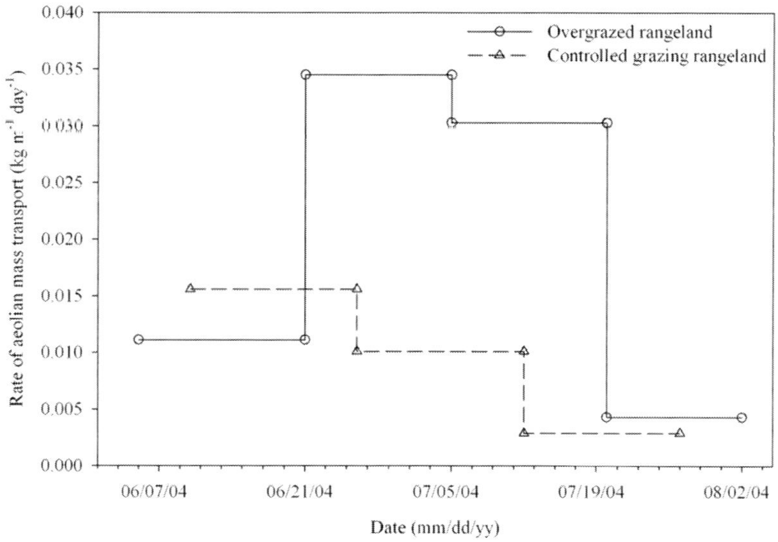

Figure 5. Mass transport rate of wind eroded aeolian sediments from (a) overgrazed rangeland site at Jaisalmer and (b) controlled grazing rangeland site at Chandan.

Carbon and Nutrient Loss from Rangelands

C and N loss through wind eroded soil from rangeland was studied at regional research station farm of CAZRI at Jaisalmer. The rangeland site was dominated with *Lasuirus sindicus* (sewan) grass. Density of sewan grass at the field site was 8000-9000 tussocks ha^{-1}. Grazing in the field site was controlled by allowing cattles to browse sewan grass to a height of 10-15 cm from desert surface. The average contents of C and N in eroded soils from the experimental rangeland site at CAZRI, Jaisalmer were 4 g kg^{-1} and 0.77 g kg^{-1}, respectively (Santra *et al.*, 2011). In general, both C and N contents were higher during dust storm events than during local wind erosion events. Overall, the C and N contents ranged from 3.01 to 5.44 g kg^{-1} and 0.30 to 0.51 g kg^{-1}, respectively. In comparison to C and N content of eroded soil mass, the average contents of C and N of surrounding top soils of the experimental site were 5.01 and 0.44 g kg^{-1}, respectively. This indicated that apart from dust emission, deposition of eroded particles in the rangelands is common, which finally led to enrichment of top soils. However, continuous overgrazing associated with severe drought situations in the desert may lead to loss of top fertile soils in quick time.

It was observed that the C and N content of eroded soil did not vary significantly with height both during dust storm and local wind erosion events ($p<0.05$) (Figure 6). However, it was observed that C and N content of eroded soil slightly increased with height during extreme wind erosion events. During local wind erosion events, C and N content did not vary with height. Carbon content of eroded soils was observed highest during dust storm event on July 9 2009 (5.13 g kg^{-1}) and then reduced to the lowest 3.35 g kg^{-1} at the end of September. Similarly, N content of eroded soils was observed highest during the dust storm event on July 9 2009 (0.50 g kg^{-1}) and then reduced to the lowest 0.33 g kg^{-1} at the end of September

(Figure 6). Month-wise comparison revealed that during the month of July, both C and N contents were significantly higher than rest periods of the year. Cumulative loss of nutrient matters along with the eroded mass during $3^1/_2$ months period from the middle of June to the end of September was 5.404 kg C ha^{-1} and 0.512 kg N ha^{-1}. The average matter loss was 0.028 kg C ha^{-1} min^{-1} during dust storm events and 0.025 kg C ha^{-1} day^{-1} during local wind erosion events (Santra *et al.*, 2011).

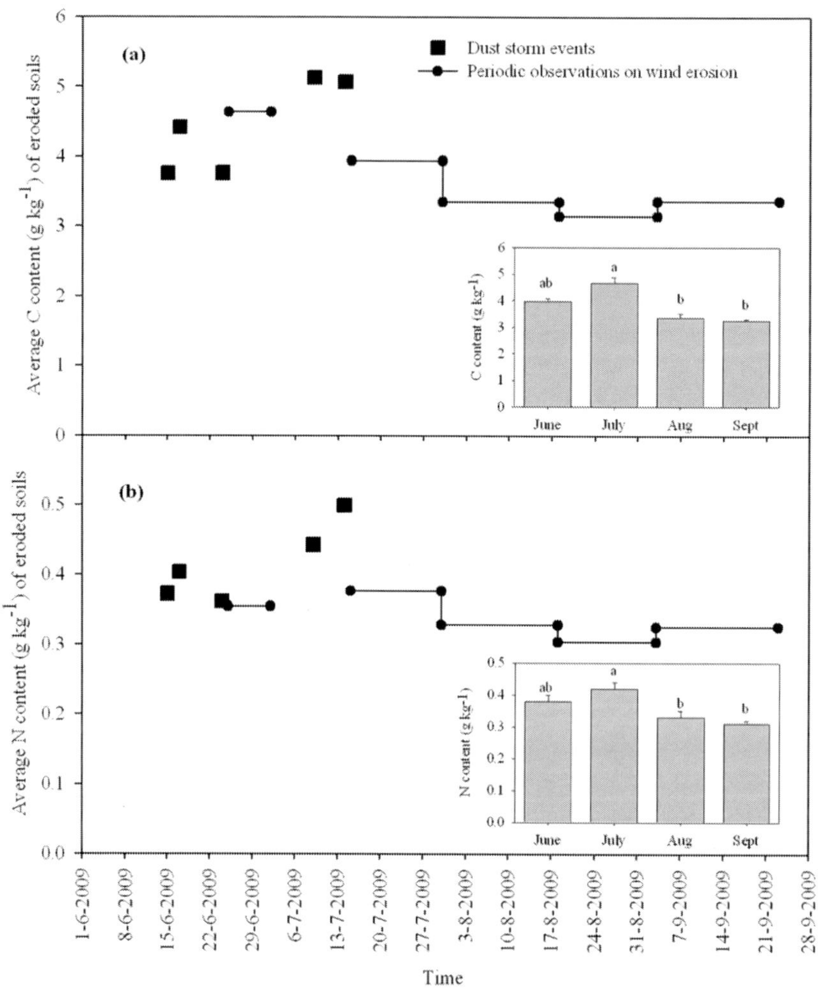

Figure 6. Periodic variation of carbon (C) and nitrogen (N) content of eroded soils; (a) C content and (b) N content Values with the same letters above the bars are not significantly different ($P<0.05$, Tukey test).

Particulate Matter (PM) in Atmosphere from Grazing Rangelands

Eroded soil mass or commonly known as dust, emitted during wind erosion events over overgrazed rangelands remains in atmosphere for a long time as suspended particles. These suspended particulate matters mix with other minute particles of the atmosphere e.g. carbon soots, smoke, salts etc. and produce a blanket of haze, which is known as aerosols. Aerosols have both direct and indirect effect on net radiative forcing and therefore play a key role in climate change of deserts as well as in the globe. Moreover, suspended particulate matters in atmosphere have an adverse effect on respiratory and cardiovascular activity of people and thus are considered as health hazard.

Particle size analysis of collected aeolian samples during dust storm event of 15-06-2009 at Jaisalmer showed that particulate matter having size less than 10 μm (PM_{10}) was 30% in eroded aeolian mass, whereas the same for desert surface was only 7.5% (Santra *et al.*, 2010). Particle size distribution of eroded soil along with the same for surface soil is presented in Figure 7. It clearly revealed that fine sized particles were more in eroded soils than on land surface and therefore have a great role to aerosol load in atmosphere.

CONCLUSIONS

Grazing based on carrying capacity aiming at 70% utilization has proved better; because during sub-normal and drought years, moderate and heavy intensities has caused heavy utilization and mortality of palatable species. The heavy intensities have shown adverse effects on palatable species on both protected and on free common grazing lands. It is therefore, important to know the type of animal to be grazed on rangeland and to decide the stocking rate on year long basis or seasonal basis to avoid any deterioration, which may otherwise lead to complete eradication of palatable perennial species. While, taking decisions on stocking rate of animals in rangeland it also should be kept in mind that the rangeland should not be overgrazed. Because, experimental studies have shown that overgrazing may lead to 3 to 5 times higher soil loss than controlled grazing due to wind erosion. Moreover, these eroded soils pose multifaceted problems in environment e.g. heavy atmospheric load of health hazardous fine dusts, loss of soil carbon and nutrients etc.

REFERENCES

Ahuja, L.D., Prajapat, M.C., Vyas, K.K., and Abbasi, G.R. (1970) Growth under different systems of grazing in *Aristia-Eleusine* pastures in arid region of the Rajasthan. *Annals of Arid Zone* 9, 266-271.

Ananth Ram, K. (1993) Carrying capacity thresholds as regulatory feed back devices for common property lands management in arid Rajasthan. In: *Proceedings of International Workshop on Common Property resources and Crisis of Pastoralism in Thar Desert.*

Bhimaya, C.P., and Ahuja, L.D. (1969) Criteria for determining condition class rangelands in Western Rajasthan. *Annals of Arid Zone* 8, 73-79.

Champion, H.G. (1936) *A preliminary survey of the forest types of India and Burma.* Indian Forest Records. Mimeo.

Crawley, M.J. (1983) *Herbivory: The dynamics of animal-plant interaction.* Blackwell scientific Publication, Oxford.

Dabadghao, P.M., and Shankarnarayana, K.A. (1973) *The grass cover of India.* ICAR, New Delhi. pp. 718.

Das, R.B., and Sarup, S. (1951) Flora in Rajputana States. *Bio-Science* 1, 36-42.

Dhir, R.P. (1977) Western Rajasthan soil their characteristics and properties. In: *Desertification and its control.* ICAR, New Delhi. pp. 102-115.

Gupta, R.K., and Saxena, S.K. (1971) Ecological studies on the protected and overgrazed rangelands in the arid zone of western Rajasthan. *Journal of Indian Botanical Society*, 50.

Kaul, K.N. (1951) A palm found from Kapurdi (Jodhpur-Rajasthan desert) *Cocos sahnii* sp. Nov. *Current Science* 20, 138.

Kumar, S., Mertia, R.S., Narita, K., Lehikawa, S., and Furukawa, A. (1999) Seasonal shifts in phenology and palatability in various grazing pressures: Implication for rangeland health. In: *Proceedings of IVth Rangelang Congress*, Australia, July19-23 Vol I, pp.550-551.

Lakhanpal, R.N., and Bose, M.N. (1951) Some tertiary leaves and fruits of Guttiferae from Rajasthan. *Journal of Indian Botanical Society* 30, 131-139.

Lakhanpal, R.N., and Singh, G. (1975) Paleobotanical history. In: *Environmental analysis of Thar desert* (Eds. R.K. Gupta and I. Prakash). English Book depot, Dehradun, pp. 38-47.

Lukose, N.G. (1984) How far the palynological evidences supports the drift of the Indian sub-continet- a case study. In: *Evolutionary Botany and Biostratigraphy* (Eds. A.K. Sharma, G.C. Mitra, and M. Banerjee). Todays and Tomorrows Punters, New Delhi. pp. 683-694.

Mertia, R.S. (1976) *Study on the floristic composition of extreme arid regions of Rajasthan.* PhD Thesis, University of Jodhpur. Rajasthan, India.

Mertia, R.S. (1992a) *Studies on improvement and utilization of rangelands of Jaisalmer region.* CAZRI Monograph No. 38, pp. 1-45.

Mertia, R.S. (1992b) Effect of grazing management on rangeland vegetation and heifers performance in western Rajasthan. *Current Agriculture* 16, 71-74.

Mertia, R.S., and Bhandari, M.M. (1992) Vegetational adaptation in the extremely arid region of the Indian desert. In: *Inter Natl. Symp. Arid Zone Research and Development.* Scientific Publisher, Jodhpur. pp. 155-164.

Mertia, R.S. (1996) Impact of grazing on the vegetation in the Indian Thar desert. In: *Towards solving the global desertification problem* (4) (Eds. T. Miyazaki and A. Tsunekawa). National Institute of Environental Studies, Japan. pp. 57-64.

Mertia, R.S., Prasad, R., Kandpal, B.K., and Narain, P. (2006) Regeneration of *Lasiurus sindicus* in relation to grazing pressure and root-zone soil moisture in arid rangelands of western Rajasthan. *Tropical Grasslands*, 40, 40-44.

Mertia, R.S., Santra, P., Kandpal, B.K., and Prasad, R. (2010) Mass-height profile and total mass transport of wind eroded aeolian sediments from rangelands of Indian Thar Desert. *Aeolian Research* 2, 135-142

Narita, K., Mertia, R.S., and Kumar, N. (1996) Vegetation coverage under different grazing pressures in arid and semi-arid areas in Thar desert. In: *Towards solving the global desertification problem* (4) (Eds. T. Miyazaki and A. Tsunekawa). National Institute of Environental Studies, Japan. pp. 49-56.

Pandey, S.C. (1964) Ecology of grasslands of Sagar, Madhya Pradesh II: A Composition of fenced associations. *Indian Journal of Botanical Society* 43, 577-605.

Paroda, R.S., Mann, H.S., and Verma, C.M. (1980) *Management of Indian arid rangelands.* CAZRI Technical Bulletin No. 4.

Ramakrishna, Y.S., Agnihotri, C.L., and Rao, R.S. (1992) Climatic resources of Thar desert and adjoining arid areas of India. In: *Perspectives of Thar and Karakum* (Eds A. Kar, R.K. Abhichandrani, K. Ananth Ram, and D.C. Joshi). Department of Sscience and Tchnology, Govt of India, New Delhi. pp. 10-20.

Roy, B.B., Dhir, R.P., and Kolarkar, A.S. (1978) *Soils of Rajasthan desert and their characteristics. In: Proceedings of Winter school on development of Rajasthan desert.* CAZRI, Jodhpur, India.

Santra, P., Mertia, R.S., Kumawat, R.N., Sinha, N.K., and Mahla, H.R. (2011) Loss of soil carbon and nitrogen through wind erosion in the Indian Thar desert. Journal of Arid Environment (*In review*).

Santra, P, Mertia, R.S., and Kushawa, H.L. (2010) A new wind erosion sampler for monitoring dust storm events in the Indian Thar desert. *Current Science* 99, 1061-1067.

Shankar, V. (1994) Vegetation of the Indian arid zone. In: *Sustainable development of the Indian Arid Zone* (Eds. R.P. Singh and Surendra Singh). Scientific Publisher, India. pp. 49-53.

Shannon, C.E., and Weaver, W. (1949) *The mathematical theory of communication.* University of Illnios Press, Urbana.

Sharma, J. P., Giri, J.D., Shyampura, R. L., and Gajbhiye, K.S. (2006) *Soil series of Gujarata.* NBSSandLUP Publication No. 120, NBSSandLUP, Nagpur, India, pp. 329.

Sharma, S.K. (1987) *Improvement and grazing management of arid rangelands at Samdari and Jodhpur.* CAZRI Monograph No. 30, pp. 39.

Shyampura, R. L., Singh, S. K., Singh, R. S., Jain, B.L., and Gajbhiye, K. S. (2002) *Soil series of Rajasthan.* NBSSandLUP Publication No. 95, NBSSandLUP, Nagpur, India, pp. 329.

Verma, C.M., and Ahuja, L.D. (1978) Growth performance of heifers under different systems of grazing on *Dichanthium-Aristida* pastures in semi-arid regions of Rajasthan. *Annals of Arid Zone* 18, 13-20.

In: Grazing Ecology: Vegetation and Soil Impact
Editor: José Ramón Arévalo
ISBN 978-1-62100-436-3

Chapter 3

THE ECOLOGY OF DOMESTIC GRAZING IN THE AUSTRALIAN ALPS

***Keith L. McDougall*[*,1], *Richard J. Williams*[2] and *Carl-Henrik Wahren*[3]**

[1]Office of Environment and Heritage, Department of Premier and Cabinet Queanbeyan, NSW, Australia
[2]CSIRO Sustainable Ecosystems, Winnellie, Australia
[3]Research Centre for Applied Alpine Ecology, Department of Agricultural Sciences, La Trobe University, Bundoora, Australia

ABSTRACT

Ungulate grazing in the Australian Alps commenced in the mid-19th Century. Prior to that, there were no large grazing animals in the alpine zone. Ecological research into the effects of domestic grazing commenced in the 1940s following concern about the effect of grazing animals on soil stability and a proposed hydro-electricity scheme. Grazing was found to increase the cover of bare ground and the risk of soil erosion. The creation of bare ground by grazing animals in grass-dominated vegetation facilitated the establishment of shrubs. In the absence of grazing, the cover of palatable forbs and shrubs increased. Cattle have had no influence on tall shrublands where there are few palatable species but they may affect the structure and function of wetland vegetation through trampling. Following a landscape scale fire in 2003, cattle were found to have had no effect on what vegetation was and was not burnt in the treeless country of the Alps. That is, fuel reduction using grazing was ineffective in fire mitigation.

Keywords: cattle, ecological processes, fire risk, trampling

---

[*] Email: keith.mcdougall@environment.nsw.gov.au, PO Box 733,Queanbeyan, NSW, 2620, Australia.

INTRODUCTION

The Australian Alps are a range of mountains with broad plateaux and deeply dissected valleys in the south-eastern part of mainland Australia (Figure 1). There are also extensive alpine environments in the island State of Tasmania. Here, we discuss aspects of the ecology and land use of the alpine environments of mainland Australia. The altitude of the Alps ranges from about 300 m a.s.l. to the highest summit, Mt Kosciuszko, at an elevation of 2228 m a.s.l. The Alps receive regular, persistent winter snow above about 1400 m. The alpine treeline ranges from 1600 m in the south to 1900 m in the north. A feature of the Alps landscape is the extensive treeless plains of plateaux valleys, where the treeline is inverted because of the accumulation of cold air in the growing season, which prevents trees on the slopes from establishing in the valleys (Williams, 1987). The vegetation of the treeless plains and alpine peaks is largely dominated by grasses and low shrubs with a high diversity of endemic inter-tussock forbs (McDougall and Walsh, 2007). Soils are low in clay, high in organic matter, acidic and nutrient poor. They are highly erodible by both water and wind following disturbance that removes vegetative cover (Costin, 1954; Williams *et al.,* 2006a).

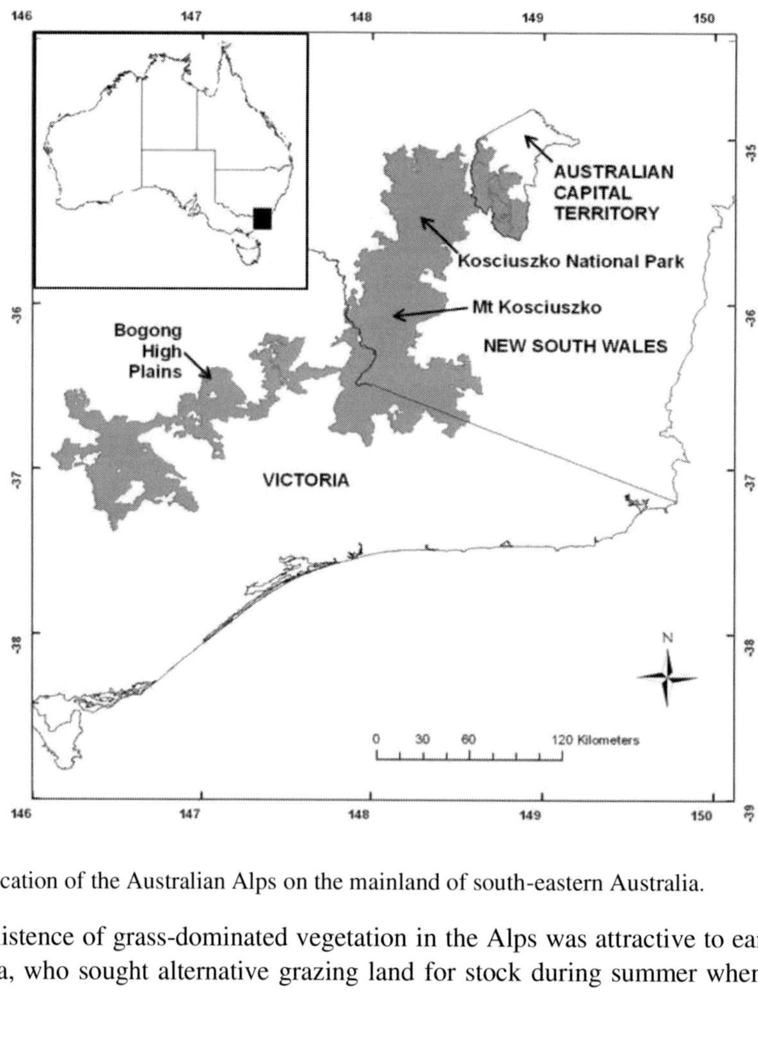

Figure 1. Location of the Australian Alps on the mainland of south-eastern Australia.

The existence of grass-dominated vegetation in the Alps was attractive to early colonists in Australia, who sought alternative grazing land for stock during summer when forage and water could be scarce in the lowlands. From the mid-19th Century large numbers of cattle and sheep were taken to the Alps each summer and then mustered and brought back to the lowlands before the snows of winter. In times of drought stock numbers could be very high (e.g. > 100,000 head in about 120 km^2 of suitable grazing land on the Bogong High Plains in the early 20th Century; Carr and Turner, 1959a).

Native vertebrate herbivores are currently absent from the mainland alpine zone and there is no evidence that they were ever abundant there. Macropods (kangaroos and wallabies) and wombats occur in the lower treeless plains and forests at low density compared with historic densities of domestic grazing animals. The former are visitors to the higher Alps and overwinter at lower elevations below the regular snowline. Wombats overwinter in the snow (Green and Osborne, 1994). The introduction of domestic stock dramatically increased grazing pressure in the Alps and changed the main grazers from soft-footed marsupials to predominantly hard-hoofed ungulates.

Grazing occurred in two State jurisdictions: New South Wales, where the principal grazing animals were sheep, and Victoria, where, in the latter half of the 20th Century, only cattle were permitted to graze. The management of grazing in the two States has differed greatly. Damage to vegetation and soils from grazing was first noted in the early 1930s and by the 1940s there was strong concern in NSW and Victoria about the effects of stock on soil stability, especially as plans were being made for a large hydroelectricity scheme. In the mid-1940s, regulatory limits were placed on grazing numbers and duration, and grazing was removed altogether from the highest summits in NSW (Costin *et al.,* 2000). In the 1950s, the stocking rate in a grassland in the Kosciuszko area was 2 – 2.5 sheep equivalent / ha (Bryant, 1973), although the overall carrying capacity for the Kosciuszko area at that time was estimated to be only about 1.2 sheep equivalent / ha (Taylor, 1956). Monitoring experiments were instigated by the governments in both States to investigate the effects of grazing on soils and vegetation (Carr and Turner, 1959a; Costin *et al.,* 2000). Domestic grazing was removed altogether from government-controlled land in NSW between the late 1940s and late 1960s before most monitoring studies were complete. In Victoria, grazing ceased in 2005 after a severe fire in 2003. Almost all of the land where domestic grazing has occurred in the Alps is now National Park, where the main aim of land management is protection of nature conservation values.

THE ECOLOGY OF DOMESTIC GRAZING IN THE AUSTRALIAN ALPS

The History of Ecological Research

Ecological research into the effect of grazing in the Australian Alps commenced in 1946 when the Victorian Government commissioned the University of Melbourne to set up exclosures on the Bogong High Plains. Two exclosures, capturing a range of plant communities, were monitored by Maisie Carr and John Turner until the 1960s (Figure 2; Carr and Turner, 1959b). After a short hiatus, the program was taken up by, in turn, the Soil Conservation Authority and La Trobe University (Wahren *et al.,* 1994), making it one of the longest running ecological monitoring projects in the world. Further monitoring plots and permanent photoquadrats were set up in the Kosciuszko area in the 1950s and 60s, as grazing was being removed (Bryant, 1969, 1973; Wimbush and Costin, 1979b, c; Scherrer and Pickering, 2005), and on the Bogong High Plains in the 1970s, where cattle continued to graze. As part of the research at Kosciuszko, sheep were re-introduced to small enclosures (Wimbush and Costin, 1979a). Since the 1980s, ecological research has focussed on the patterns and processes of grazing effects; for example, diet and behaviour of cattle, shrub – grass dynamics, influence on fire patterns, regeneration after fire, and interaction with predicted climate change.

Figure 2. This grazing exclosure at Pretty Valley on the Bogong High Plains was fenced in 1947. By 2007, when this photo was taken, there had been a dramatic increase in the cover of palatable forbs and shrubs.

Because grazing commenced almost a century before ecological research and no areas are known to have escaped grazing, ecological studies have involved measuring (i) vegetation in exclosures and adjoining grazed areas, (ii) the response of vegetation following removal of stock or (iii) the effects of re-introducing stock (e.g. in enclosures). Although it has not been possible to investigate the effects of introducing stock to the Australian Alps, the long duration of ecological studies has provided invaluable information to land managers.

Diet and Behaviour of Cattle

The behaviour and diet of cattle in the Alps have been investigated by observation, faecal-examination and oesophageal-fistulation techniques (Van Rees and Hutson, 1983; Van Rees, 1984; Van Rees and Holmes, 1986). In treeless areas, cattle were found to prefer grassland and low heathland for grazing and resting. Tall shrublands were rarely visited. Wetlands were visited mainly for water.

Cattle are selective grazers in the Alps. The principal species eaten by cattle are common grasses (*Austrodanthonia nudiflora, Poa costiniana, P. hiemata* and *P. hothamensis*), sedges

(*Carex* spp.), forbs (e.g. *Asperula gunnii, Celmisia* spp. and *Leptorhynchos squamatus*), and low shrubs (*Asterolasia trymalioides* and *Grevillea australis*). Tall shrubs are not consumed. Grasses are preferred at the beginning of the grazing season (making up about half of the live dry matter in extrusa samples), grasses and shrubs are preferred in the middle of the grazing season (about 25% and 50% live dry matter respectively), and shrubs and forbs are eaten towards the end of the season (about 30% and 50% live dry matter respectively). The digestibility and nitrogen content of the majority of species eaten by cattle declines during the grazing season. By the time cattle are taken back to lowland pasture, the energy and nitrogen content of plant species in the diet are suboptimal for maintenance of cattle live weight (Van Rees, 1984; Van Rees and Holmes, 1986).

Plant Species Composition

Long-term exclusion plots established on the Bogong High Plains in the 1940s have shown clearly that, in the absence of grazing, the abundance of many palatable species increases substantially. This is particularly so for the major forbs and low shrub *Asterolasia trymalioides* (Carr and Turner, 1959b; Wahren *et al.,* 1994), the cover of which changed little on grazed plots. In contrast, most shrubs in both open and closed heathland, such as *Hovea longifolia, Prostanthera cuneata, Phebalium squamulosum* and *Orites lancifolia* are non-palatable. Long-term monitoring in reference areas where these shrubs predominate has shown that their cover is unaffected by grazing (Wahren, 1997; Wahren *et al.,* 1994).

The transects and photoquadrats established in the Kosciuszko area after the removal of grazing initially showed an increase in the cover of palatable species (forbs and the shrub *Grevillea australis*) and a decrease in the cover of bare ground (Wimbush and Costin, 1979b,c; Scherrer and Pickering, 2005), as had been found in the Bogong High Plains studies. However, over the entire period of measurement (1959 – 2001), the cover of most growth forms and bare ground fluctuated, a response attributed in part to periodic drought and the spatial variability of the vegetation (Scherrer, 2004).

In experimental trials at Kosciuszko where sheep were re-introduced in enclosures (at a stocking rate of about 5 sheep / ha), the cover of some palatable forbs (e.g. *Aciphylla simplicifolia, Craspedia* spp.) increased in the absence of grazing and the cover of unpalatable shrubs (*Bossiaea foliosa* and *Oxylobium ellipticum*) increased on grazed plots. The cover of bare ground decreased on ungrazed plots. Once grazing had been removed, the biomass of intertussock forbs increased from 0.3 kg/ha to 1 kg/ha over a period of 10 years (Wimbush and Costin, 1979a).

Within stock exclosures in the subalpine plains in the Kosciuszko area (Bryant, 1969) there was a rapid increase in the number of tree seedlings establishing and persisting, suggesting that grazing will inhibit movement of the treeline expected to occur with climate change. There was an increase in shrub cover (*Bossiaea foliosa, Hovea montana* and *Pimelea linifolia*) of 15 – 30% in the exclosures over the first three years but the cover of these species declined thereafter. In grassland areas within exclosures there was an increase in the cover of forbs where there were abundant inter-tussock spaces but little change in species composition where grass cover was dense.

Bryant (1973) used line transects to investigate changes in species composition following the re-introduction of sheep and cattle grazing after four years without grazing. In dense grassland, the cover of forbs (especially *Geranium antrorsum* and *Ranunculus graniticola*) and one shrub species (*Hovea montana*) fell quickly although there was little change in overall cover during the decade of measurement. In sparsely-vegetated grassland, the cover of mat-forming shrubs (*Pultenaea* spp.) increased but overall vegetative cover decreased.

The selectivity of domestic grazing animals in the Australian Alps is thought to have pushed some native plant species to near extinction. The forb *Ranunculus anemoneus* is apparently highly palatable. It was once widely distributed in the Kosciuszko area but by the 1960s was becoming difficult to find and was believed to be a species of rock ledges (Costin *et al.,* 2000). Now, in the long-absence of grazing, it is found in a wide range of habitats and is reasonably common. The tall grass *Chionochloa frigida* has made a similar return from obscurity in the Kosciuszko area.

Ecological Processes

Cattle may alter the structure of communities in the Alps through trampling and selective grazing. The processes leading to structural change have been well studied in heathland and grassland communities but are largely inferred in other communities.

Shrub invasion has been recorded in many grasslands in the Alps both in areas grazed throughout the 20th century and in areas where grazing was withdrawn in the mid 20th century (e.g. Lawrence, 1999; McDougall, 2003; McDougall, unpublished data). The key to understanding this process is the interaction between disturbance regimes and the life-histories of the shrubs (Williams and Ashton, 1987, 1988; Williams, 1990). Shrub seedlings require bare ground to establish and grazing manifestly increases the abundance of bare ground. Where non-palatable shrubs have established (e.g. *Bossiaea foliosa* and *Hakea microcarpa*), they have invariably increased in cover because they are not browsed and also resprout vigorously from dormant buds after trampling. Such shrubs can therefore persist following establishment. In contrast, where palatable shrubs such as *Asterolasia trymalioides* have established, their growth is inhibited by grazing because they are both browsed and, due to the absence of dormant buds, incapable of resprouting (Wahren *et al.,* 1994). Therefore such shrubs remain suppressed (Figure 3).

If shrubs establish and cattle grazing is subsequently discontinued, the cover of some palatable shrubs may increase substantially, as has happened with *Asterolasia trymalioides* inside an exclosure fenced in 1947 (Carr and Turner, 1959b; Wahren *et al.,* 1994), and on experimentally induced patches of bare ground (Williams and Ashton, 1987). Whilst the cover of such palatable shrubs may increase in the 20-50 years following the removal of cattle, these shrubs do not persist. There are two reasons for this. First, they are subject to replacement by forbs and grasses as they senesce and die (Williams and Ashton, 1988; Williams, 1990; Wahren *et al.,* 1994). Second, they will only establish and regenerate by seed (Williams and Ashton, 1987), for which they require patches of bare ground, which is minimal (< 3%) in areas long-ungrazed by cattle (Wahren *et al.,* 1994).

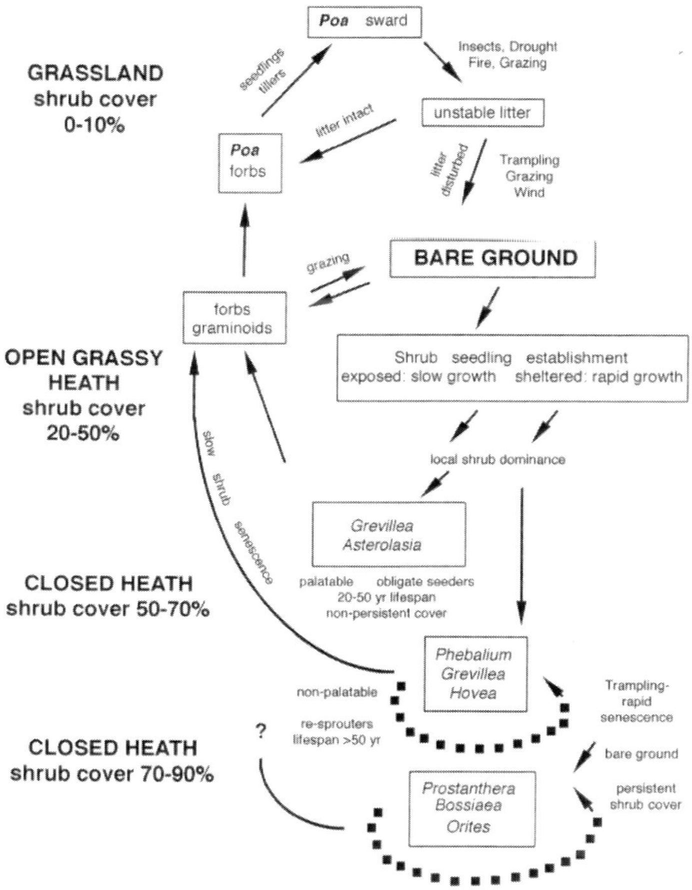

Figure 3. Summary of the dynamics of the heath-grass complex (closed heath, open heath, grassland) on well-drained alpine humus soils on the Bogong High Plains. Taxa: grasses - *Poa hiemata* (Snow Grass). Shrubs - *Grevillea australis* (Alpine Grevillea), *Asterolasia trymalioides* (Alpine Star Bush), *Phebalium squamulosum* (Forest Phebalium), *Hovea longifolia* (Rusty Pods), *Prostanthera cuneata* (Mint Bush), *Bossiaea foliosa* (Mountain Bossiaea) and *Orites lancifolia* (Alpine Orites).

The increase in shrub cover in the Alps where grazing has been withdrawn has commonly been attributed to the withdrawal of grazing. For example, a management plan for an historic homestead, reported that "long-term residents of the region note that the native vegetation on the plain is noticeably different to when they were being intensively grazed. Most obvious are the increased patches of *Hakea microcarpa*." (Jill Sheppard Heritage Consultants, 2003). The withdrawal of grazing has probably released seedlings from trampling and grazing pressure, and allowed their expansion as has been observed. However, the primary reason for the increase is likely to be the creation of bare ground when the vegetation was grazed. For this reason, shrubs are increasing in cover both in areas with continuous grazing and where grazing has been withdrawn.

The treeline dominant (*Eucalyptus pauciflora* sens. lat.) has also invaded grassland over the past 50 years in the Kosciuszko region following the cessation of grazing. The foliage of this tree is browsed by sheep. Seedlings may be suppressed by grazing (Costin *et al.*, 1959). The encroachment of trees and expansion of shrubs are expected consequences of climate change in mountains worldwide (e.g. Sturm *et al.*, 2005; Tape *et al.*, 2006; Blok *et al.*, 2010, Hallinger *et al.*, 2010) and invasion of subalpine grassland by *Eucalyptus pauciflora* in the past 30 years has been documented near Mt Hotham in the Australian Alps (Wearne and Morgan, 2001). Climate change may act with land use change to determine treeline responses to global warming (Gehrig-Fasel *et al.*, 2007) hence an unexpected consequence of grazing in the Australian Alps could be that it will be difficult or impossible to distinguish between the effects of climate change and the lasting effects of grazing.

Although wetland communities are known to be readily damaged by trampling because both the peat soils and dominant cover plant (*Sphagnum cristatum*) are relatively soft, in Australia there have been no studies of the direct effect of trampling on wetland community structure. This is probably because much of the trampling damage had already occurred when ecological research began in the Australian Alps and it was then only possible to investigate the effect of removing stock. Wetland areas excluded from cattle on the Bogong High Plains in 1980 showed a substantially greater rate of recolonization of exposed soil than areas to which cattle had continued access (Wahren *et al.*, 2001b). In an exclosure fenced in 1946, the cover of *Sphagnum cristatum* increased greatly between 1979 and 1999 but did not change in similar wetlands adjoining the exclosure or elsewhere in the vicinity (McDougall, 2007). The ungrazed wetland has isolated pools, relatively deep peats (> 1 m), *Sphagnum* mounds and impeded drainage not observed elsewhere in this area (Tolsma *et al.*, 2005). In the Kosciuszko area, two palatable species, *Empodisma minus* and *Carex gaudichaudiana* increased in cover in a wetland following the removal of grazing. The impedance of water flow by increasing species such as this is thought to have facilitated the spread of *Sphagnum cristatum* and general recovery of the wetland vegetation (Wimbush and Costin, 1979b, c).

In herb-dominated communities on steep lee slopes where soils remain moist until well into the growing season, cattle grazing and trampling can increase the amount of bare ground substantially, from less than 2% to more than 20%; (Wahren, 1997; Wahren *et al.*, 2001a). This increases the risk of soil erosion.

Domestic stock interfere with natural insect – plant interactions in the Australian Alps. Native invertebrates (swift moths and case moths) eat and may kill grass tussocks but the remaining plant litter usually remains attached to both the parent tussock and to the ground (Carr and Turner, 1959a). This litter is highly susceptible to disturbance by trampling, which can dislodge the litter. The litter may then be blown or washed away, creating bare ground. In ungrazed sites, litter is far more persistent and the amount of bare ground resulting from insect attack is 5-10 times less than that in comparable sites grazed by cattle (Williams and Ashton, 1987; Wahren *et al.*, 1994). As described above, the creation of bare ground may favour shrub colonisation rather than recolonisation by grasses.

Fire Regimes

Two landscape scale fires burnt most of the Australian Alps in the last century (in 1939 and 2003; Williams *et al.*, 2006b, 2008). Until the mid 20th Century, much of the grazed area was also burnt at low intensity by graziers to promote new grass growth and suppress woody species (Costin *et al.*, 2000). The evidence from fire scars on trees and dendrochronology indicates, however, that fire was relatively rare in pre-European times (Banks, 1989; Zylstra, 2006). There is no evidence that native Australian peoples practiced 'fire stick farming' in the high parts of the Alps and there would have been little reason to do so, as the target game species were uncommon or absent (Costin *et al.,* 2000). Despite this, there is a widely held, albeit erroneous view that the removal of grazing from most of the Alps has resulted in an unnatural fire regime because of the decrease in low intensity deliberate fire and, moreover, an increased risk of fire because of the reduced consumption of fuel by stock (e.g. Jurskis *et al.,* 2006).

The 2003 fire, which burnt more than half of the Alps, gave an opportunity to test whether grazed areas burnt less than long-ungrazed areas. Williams et al. (2006b) found no statistically significant difference between grazed and ungrazed areas in the proportion of about 400 randomly located sample sites in burnt vegetation. In shrubby communities, grazing did not significantly lower the severity of fire, as measured by the diameter of burnt twigs. The lack of a grazing effect was attributed to shrub dynamics, cattle diet and behaviour (as described above), and fuel flammability. The most flammable communities are dry with high biomass dominated by woody species. These often also occur on the steeper slopes where the rate of fire spread, and therefore fire intensity, is greater than on gentle slopes where grasslands with low biomass predominate.

In the fires of 2003, it is clear that, at the landscape scale, grazing had no effect on what was and wasn't burnt in the treeless country of the Alps. Grazed and long-ungrazed country burnt in some places and not others. What was burnt was a function of fire behaviour in surrounding forests, where ignition began from lightning strikes, rather than the nature of the mountain tops. Most of the Alps was burnt in 1939 when all of the area was grazed.

Soils

The exposure of soil through grazing and the increased risk of erosion of exposed soil through the action of needle ice, water and wind were identified in many of the studies described above. However, there appear to have been no studies comparing the rate of soil loss in grazed and ungrazed situations in the Alps. This is curious considering that the initial reason for conducting ecological research was to investigate soil stability. Costin et al. (1960), however, found that run-off and soil loss in forested and treeless communities after rain were negatively correlated with percentage ground cover. Any activity, such as grazing, that leads to an increase in bare ground will therefore cause enhanced soil loss. Bryant (1973) used aerial photography to assess changes in erosion severity and vegetation cover between 1944 and 1970 in a subalpine grassland, which was rested from grazing between 1958 and 1962. The proportion of the area with severe erosion increased with time as did the proportion of the area with complete vegetative cover. This suggests that the reduced stocking rate associated with grazing controls introduced in 1944 allowed recovery of well vegetated land but, once large areas of soil were exposed, these would continue to erode regardless of stocking rate.

In the Kosciuszko alpine area, soil erosion was attributed to grazing before research had associated grazing with exposure of bare ground, and extensive and costly remediation works were undertaken to repair damage (e.g. Clothier and Condon, 1968; Bryant, 1971).

Fauna

The impact of domestic grazing on native fauna is largely unknown. Domestic grazing animals are regarded as a threat to aquatic invertebrates in the Alps, many of which are highly localised (e.g. a stonefly, *Thaumatoperla alpina*; McKay *et al.,* 2005), because of potential changes to wetland and stream structure caused by trampling.

CONCLUSION

Grazing research in the Australian Alps has passed through three phases since its inception. In the first – the hydroelectricity phase – the aim of research was to determine the impact of grazing on vegetation cover and soil stability. In the second – the nature conservation phase – the aim of research was to determine the impact of grazing on biodiversity values. In the third and current – the fire and fuel phase – the aim has shifted to the putative benefit of grazing in reducing the risk of fire in the Alps (Anon, 2003). This phase has arisen in a decade of major wildfires that have destroyed life and property. These three generalised phases are not independent, however.

In the first phase, ecological research demonstrated that grazing by domestic animals results in greater exposure of soil and an increased risk of erosion. Dry communities do appear to have some resilience to grazing as shown by the increase in cover of palatable forbs and shrubs in exclosures. Importantly, however, there appears to be a degree of soil exposure beyond which the removal of grazing will not result in rapid natural recovery of vegetative cover. Large eroded areas in the Kosciuszko alpine zone in the 1940s continued to erode long after stock were removed and were extremely costly to repair. Trampling in wetlands was also found to lead to long term damage requiring remediation. In the second phase, the interaction between grazing and the processes of shrub and herb establishment were investigated. The creation of bare ground was found to favour shrub establishment, although continued grazing and trampling could also inhibit shrub invasion. Overall, shrubs appear to be prevailing regardless of grazing. Cattle were found to be selective grazers, preferring nutrient-rich forbs that grow in spaces between grass tussocks. An increased knowledge of Alps biota in this phase has identified species at risk from trampling and the subsequent alteration of drainage processes in wetlands. The clear and compelling conclusion from this phase of research is that livestock grazing is a threat to most soil, water and nature conservation values in alpine landscapes in Australia.

The current phase of research is ill-conceived because there is no evidence that grazing in the Alps can beneficially influence fire patterns at broad scales and the only evidence is, in fact, to the contrary. Given the number of catastrophic fires in the past decade, however, it is unsurprising that people are searching for means to reduce fire risk. The 'alpine grazing reduces blazing' paradigm is based on the assumption that fire extent and severity can be mitigated by grazing, because fuel loads are reduced. As mentioned above, this paradigm was shown to be unsupported across the Bogong High Plains following the 2003 fires; whatever fuel reduction there may have been did not translate into a reduction of extent or severity of fire in grazed compared with ungrazed areas (Williams *et al.,* 2006b). Grazing is also highly unlikely to reduce fire risk in forested environments in the Australian Alps for two reasons.

First, the dominant fuels in these woodlands and forests are leaf, twig and bark litter from the trees and understorey, which are, by and large, not palatable to livestock. Second, the effects of fuel reduction on mitigation of area burnt in the ecosystems of south-eastern Australia are highly contingent on fire weather at the time of burning (Bradstock, 2010; Price and Bradstock, 2010). Thus, when fire weather is severe any minor-modest reduction in fine fuel loads as a consequence of grazing is highly likely to be ineffective in reducing spread or intensity at the landscape scale. Research from the first two phases demonstrated that grazing affects species composition, vegetative cover and community structure. The task for land managers now is to weigh up the risks to biodiversity and land stability from grazing against the putative benefits of grazing to fire management.

References

Anon. (2003). A Nation Charred: Report on the inquiry into bushfires. House of Representatives Select Committee into the recent Australian bushfires, 23 October 2003. The Parliament of the Commonwealth of Australia, Canberra.

Banks, J. C. (1989). A history of forest fire in the Australian Alps. In: R. B. Good (Ed.), *The Scientific Significance of the Australian Alps*. Australian Alps Liaison Committee and Australian Academy of Science. Canberra.

Blok, D., Heijmans, M. M. P. D., Schaepman-Strub, G., Kononov, A. V., Maximov, T. C., and Berendse, F. (2010). Shrub expansion may reduce summer permafrost thaw in Siberian tundra. *Global Change Biology, 16,* 1296-1305.

Bradstock, R. A. (2010). A biogeographic model of fire regimes in Australia: current and future implications. *Global Ecology and Biogeography, 19,* 145-158.

Bryant, W. G. (1969). Vegetation and ground cover trends following the exclusion of stock at three sites in the Snowy Mountains, New South Wales. *Journal of the Soil Conservation Service of New South Wales, 25,* 183-198.

Bryant, W. G. (1971). Deterioration of vegetation and erosion in the Guthega catchment area, Snowy Mountains, New South Wales. *Journal of the Soil Conservation Service of New South Wales, 27,* 62-81.

Bryant, W. G. (1973). The effect of grazing and burning on a mountain grassland, Snowy Mountains, New South Wales. *Journal of the Soil Conservation Service of New South Wales, 29,* 29-44.

Carr, S. G. M., and Turner, J. S. (1959a). The ecology of the Bogong High Plains I. The environmental factors and the grassland communities. *Australian Journal of Botany, 7,* 12-33.

Carr, S. G. M., and Turner, J. S. (1959b). The ecology of the Bogong High Plains II. Fencing in grassland C. *Australian Journal of Botany, 7,* 34-63.

Clothier, D. P., and Condon, R. W. (1968). Soil conservation in alpine catchments. *Journal of the Soil Conservation Service of New South Wales, 24,* 96-113.

Costin, A. B. (1954). *A Study of the Ecosystems of the Monaro Region of New South Wales*. Government Printer, Sydney.

Costin, A. B., Gray, M., Totterdell, C. J., and Wimbush D. J. (2000). *Kosciuszko Alpine Flora*. CSIRO Publishing: Collingwood.

Costin, A. B., Wimbush, D. J., Kerr, D., and Gay, L. W. (1959). Studies in catchment hydrology in the Australian Alps. I. Trends in soils and vegetation. Division of Plant Industry Technical Paper No. 13. CSIRO, Melbourne.

Costin, A. B., Wimbush, D. J. and Kerr, D. (1960). Studies in catchment hydrology in the Australian Alps. II. Surface run-off and soil loss. Division of Plant Industry Technical Paper No. 14. CSIRO, Melbourne.

Gehrig-Fasel, J., Guisan, A., and Zimmermann, N. E. (2007). Tree line shifts in the Swiss Alps: climate change or land abandonment? *Journal of Vegetation Science, 18,* 571-582.

Green, K., and Osborne, W. S. (1994). *Wildlife of the Australian Snow-Country*. Reed: Sydney.

Hallinger, M., Manthey, M., and Wilmking, M. (2010). Establishing a missing link: warm summers and winter snow cover promote shrub expansion into alpine tundra in Scandinavia. *New Phytologist, 186,* 890-899.

Jill Sheppard Heritage Consultants (2003). Currango Historic Precinct Conservation Management and Interpretation Plan. NSW National Parks and Wildlife Service, South West Slopes Region and Cultural Heritage Division, Tumut, Australia.

Jurskis, V., de Mar, P., and Aitchison, B. (2006). Fire management in the Alpine region. 10th Biennial Australasian Bushfire Conference, Brisbane, June 2006. Paper accessed at http://www.griffith.edu.au/conference/bushfire2006/ on the 17th of April, 2011.

Lawrence, R. E. (1999) Vegetation changes on the Bogong High Plains from the 1850s to 1950s. *Transactions of the Royal Society of Victoria, 111,* xxix – lii.

McDougall, K. L. (2003). Aerial photographic interpretation of vegetation changes on the Bogong High Plains, Victoria, between 1936 and 1980. *Australian Journal of Botany 51,* 251-256.

McDougall, K. L. (2007). Grazing and fire in two subalpine peatlands. *Australian Journal of Botany, 55,* 42-47.

McDougall, K. L., and Walsh, N. G. (2007). Treeless vegetation of the Australian Alps. *Cunninghamia, 10,* 1-57.

McKay, S., Bryce, C., and Papas, P. (2005). *Impacts of fire on the distribution of a predatory stonefly (Eustheniidae: Thaumatoperla alpina) in the Bogong alpine region*. Freshwater Ecology, Arthur Rylah Institute for Environmental Research, Technical Report Series No.155, Department of Sustainability and Environment, Melbourne; La Trobe University, Wodonga, Victoria.

Price, O. F., and Bradstock, R. A. (2011). Quantifying the influence of fuel age and weather on the annual extent of unplanned fires in the Sydney region of Australia. *International Journal of Wildland Fire, 20,* 142-151.

Scherrer, P. (2004). Monitoring vegetation change in the Kosciuszko alpine zone, Australia. PhD Thesis, School of Environmental and Applied Sciences, Griffith University, Gold Coast, Australia.

Scherrer, P., and Pickering, C. P. (2005). Recovery of alpine vegetation from grazing and drought: data from long-term photoquadrats in Kosciuszko National Park, Australia. *Arctic, Antarctic, and Alpine Research, 37,* 574-584.

Sturm, M., Schimel, J., Michaelson, G., Welker, J. M., Oberbauer, S. F., Liston, G. E., Fahnestock, J., and Romanovsky, V. E. (2005). Winter biological processes could help convert arctic tundra to shrubland. *BioScience, 55,* 17-26.

Tape, K., Sturm, M., and Racine, C. (2006). The evidence for shrub expansion in northern Alaska and the pan-Arctic. *Global Change Biology, 12,* 686-702.

Taylor, A. C. (1956). Snow lease management. *Journal of the Soil Conservation Service of New South Wales, 12,* 33-43.

Tolsma, A., Shannon, J., Papst, W. A., Rowe, K., and Rosengren, N. J. (2005). *An assessment of the condition of mossbeds on the Bogong High Plains.* Report to the Victorian Department of Sustainability and Environment. Arthur Rylah Institute for Environmental Research and Research Centre for Applied Alpine Ecology, La Trobe University, Melbourne.

Van Rees, H. (1984). *Behaviour and diet of free-ranging cattle on the Bogong High Plains,* Victoria. Environmental Studies Publication 409. Department of Conservation, Forests and Lands, Melbourne.

Van Rees, H., and Holmes, J. H. G. (1986). The botanical composition of the diet of free-ranging cattle on an alpine range in Australia. *Journal of Range Management, 39,* 392-395.

Van Rees, H., and Hutson, G. D. (1983). The behaviour of free-ranging cattle on an alpine range in Australia. *Journal of Range Management, 36,* 740-743.

Wahren, C-H. A. (1997). *Vegetation dynamics on the Bogong High Plains.* Ph.D. thesis, Monash University: Melbourne.

Wahren, C-H., Papst, W.A., and Williams, R.J. (1994). Long-term vegetation change in relation to cattle grazing in subalpine grassland and heathland on the Bogong High Plains: an analysis of vegetation records from 1945 to 1994. *Australian Journal of Botany, 42,* 607-639.

Wahren, C-H., Williams, R. J. and Papst, W. A. (2001a). Alpine and subalpine snow patch vegetation on the Bogong High Plains, southeastern Australia. *Journal of Vegetation Science, 12,* 779-790.

Wahren, C-H., Williams, R. J. and Papst, W. A. (2001b). Vegetation change and ecological processes in alpine and subalpine *Sphagnum* bogs of the Bogong High Plains, Victoria, Australia. *Arctic, Antarctic and Alpine Research, 33,* 357-68.

Wearne, L. J., and Morgan, J. W. (2001). Recent forest encroachment into subalpine grasslands near Mount Hotham, Victoria, Australia. *Arctic, Antarctic and Alpine Research, 33,* 369-377.

Williams, R. J. (1987). Patterns of air temperature and accumulation of snow in subalpine heathlands and grasslands on the Bogong High Plains, Victoria. *Australian Journal of Ecology, 12,* 153-163.

Williams, R. J. (1990). Growth of subalpine shrubs and snow grass following a rare occurrence of frost and drought in south-eastern Australia. *Arctic and Alpine Research, 22,* 412-422.

Williams, R. J., and Ashton, D. H. (1987). The effects of disturbance and grazing by cattle on the dynamics of heathland and grassland communities on the Bogong High Plains, Victoria. *Australian Journal of Botany, 35,* 413-31.

Williams, R. J., and Ashton, D. H. (1988). Cyclical patterns of regeneration in subalpine heathland communities on the Bogong High Plains. Victoria. *Australian Journal of Botany, 36,* 605-619.

Williams, R. J., McDougall, K. L., Wahren, C-H., Mansergh, I. M., Rosengren, N. J., and Papst, W. A. (2006a). Alpine Landscapes. In P. Attiwill and B. Wilson (Eds.), Ecology: An Australian Perspective (pp 557- 572). Oxford University Press.

Williams, R. J., Wahren, C-H., Bradstock, R. A. and Müller, W. J. (2006b). Does alpine grazing reduce blazing? A landscape test of a widely-held hypothesis. *Austral Ecology, 31,* 925-936.

Williams, R. J., Wahren, C-H., Tolsma, A. D., Sanecki, G. M., Papst, W. A., Myers, B. A., McDougall, K. L., Heinze, D. A., and Green, K. (2008). Large fires in Australian alpine landscapes: their part in the historical fire regime and their impacts on alpine biodiversity. *International Journal of Wildland Fire, 17,* 793-808.

Wimbush, D. J., and Costin, A. B. (1979a). Trends in vegetation at Kosciusko. I. Grazing trials in the subalpine Zone, 1957-1971. *Australian Journal of Botany, 27,* 741-87.

Wimbush, D. J., and Costin, A. B. (1979b). Trends in vegetation at Kosciusko. 11. Subalpine range transects, 1959-1978. *Australian Journal of Botany, 27,* 789-831.

Wimbush, D. J., and Costin, A. B. (1979c). Trends in vegetation at Kosciusko III. Alpine range transects, 1959-1978. *Australian Journal of Botany, 27,* 833-871.

Zylstra, P. (2006). *Fire history of the Australian Alps. Prehistory to 2003.* Australian Alps Liaison Committee, Canberra.

In: Grazing Ecology: Vegetation and Soil Impact
Editor: José Ramón Arévalo
ISBN 978-1-62100-436-3

Chapter 4

GRAZING MANAGEMENT AND IMPACT IN THE CANARY ISLANDS: RETHINKING SUSTAINABLE USE

José Ramón Arévalo[1]*, Silvia Fernández-Lugo[1], Lea de Nascimento[1], Luis Alberto Bermejo[2] and Agustín Naranjo[3]

[1] Department of Ecology, Faculty of Biology, University of La Laguna. Avda, Astrofísico Francisco Sánchez, La Laguna (Tenerife), Spain

[2]Department of Agricultural Engineering, Production and Economy, Technical School of Agricultural Engineering, University of La Laguna, Ctra. de Geneto, La Laguna (Tenerife), Spain

[3] Department of Geography, Faculty of Geography and History, University of Las Palmas de Gran Canaria, Las Palmas de Gran Canaria (Gran Canaria), Spain

ABSTRACT

On an oceanic island, the positive relationship between livestock management and biodiversity conservation is not as clear as on the mainland. In the sensitive island's ecosystems, introduced herbivores have been considered to cause devastating effects on vegetation, and be responsible for overgrazing, ecosystem degradation, and biodiversity loss. Eradication of introduced herbivores has been proposed as a conservation tool in some islands. However, this proposal needs to be carefully considered, since there are complex and established ecological interactions between native and introduced species.

The goal of this chapter is to discuss the livestock management for biodiversity conservation, with special attention on the Canary Islands (Spain) as an example. We present a review of the research done on this topic in the Canaries, looking for general conclusions and explanations of the conflicting results. Much of the research is restricted in area, however, consistent works are presented that can help us to understand the answer of the environment to grazing impact on these islands.

Although some negative impacts of grazing were revealed, in general they have been restricted to some specific areas or species. In those cases where goat grazing is threatening endemic and endangered populations, control of goat densities, grazing

* Author for correspondence (Email: jarevalo@ull.es)

restrictions and use of fences should be encourage, however drastic measures as eradication of goats are not ecologically or socially feasible. On the other hand, grazing of small goat herds on traditionally managed ecosystems seems to be a sustainable activity, which removal have been related with a reduction in the number of species, changes in species composition and also an increase in exotic species.

Keywords: Canary Islands, conservation, eradication, grazing management, goats

INTRODUCTION

Grazing is one of the most important management activities worldwide. This has allowed for the development of an extensive bibliography on the responses of plant communities to grazing. However, we are still far away from a consensus regarding the effects of grazer animals on ecosystems. Grazing effects on species richness and composition are contradictory in many studies (Olff and Ritchie, 1998). Researchers have documented that herbivores can enhance, have a negative impact, or have weak or even no effects on species diversity and plant community composition (e.g. Olff and Ritchie, 1998; Stohlgren *et al.,* 1999; Osem *et al.,* 2002; Casado *et al.,* 2004; de Bello *et al.,* 2007), as well as, a variety of effects on soil nutrients (McIntosh *et al.,* 1997; Bakker *et al.,* 2004; Peco *et al.,* 2006; Fernández-Lugo *et al.,* 2009a). The lack of consistent responses of vegetation to grazing, confirmed in a range of research studies, has been attributed to evolutionary history of grazing, productivity gradients and grazing intensity (Milchunas and Lauenroth, 1993). Herbivore type and size has also been proven to differently shape plant community responses to grazing (Olff and Ritchie, 1998). Other causes related to these contradictory results are the differences in the amount of plant and animal species biodiversity (Sanderson *et al.,* 2004; Clergue *et al.,* 2005).

A comprehensive analysis of all the factors described above is necessary to understand the impact of grazing, and its positive or negative effects on ecological processes such as regeneration, diversity, biomass production, etc. The effects of different environmental gradients or human activity (i.e. altitude, precipitation, temperature, management regime, etc.), should also be taken into account in order to isolate the impact of grazing on plant communities. These environmental variables are in many studies more important than grazing to explain species diversity and composition, e.g. altitude (Brinkmann *et al.,* 2009; Zhang and Dong, 2009), precipitation (Milchunas *et al.,* 1988), productivity (Osem *et al.,* 2004), specific plant community (Briske and Noy-Meir, 1989) or soil nutrient content (Fernández-Lugo *et al.,* 2009a). Consequently, grazing impact on plant communities cannot be evaluated on its own. Although it is an important element in shaping plant communities, there are a multitude of factors (as above mentioned) that interact with grazing, leading to a variety of effects on ecosystems. In addition, other factors related to methodology can also affect the result of a research. For instance, the scale of sampling appears to play a significant role in conditioning the response of plant communities to grazing (Olff and Ritchie, 1998; Fuhlendorf and Smeins, 1999).

Numerous scenarios can be found in managed ecosystems, as a result of the interaction of the different factors previously mentioned. It is common to find discordant discourse about the compatibility between livestock grazing and diversity conservation. There are those in the scientific community who defend the removal of grazing as a way to ensure conservation of ecosystems (e.g. Hill *et al.,* 1998; Fraser and Chisholm, 2000), as well as defenders of the maintenance of this activity as the only way to conserve agroecosystem diversity (e.g. Perevolotsky and Seligman, 1998; Peco *et al.,* 2006), and the cultural, social, and economic values that these traditional activities support. Awareness of biodiversity conservation has burgeoned in the past two decades (Wilcove, 1993; Baydack and Campa, 1999; Millennium Ecosystem Assessment, 2005). While the initial, and almost exclusive, focus of conservation was on preserves, hot spots and natural ecosystems (Eken *et al.,* 2004), over the past several years, interest has expanded to include areas managed for agricultural production (Krebs *et al.,* 1999; Benton *et al.,* 2003). This has resulted in a noticeable increase of the literature on farmland biodiversity (e.g. Aviron *et al.,* 2009; Swagemakers *et al.,* 2009). The inclusion of lands grazed by domestic herbivores in the analysis of biodiversity conservation has lead to an interesting paradigm shift in managing ecosystems. In the last century common and habitual management practices led to the homogenization of the farmland landscapes worldwide (Jongman, 2002), and recent works suggest that homogeneous land management is one of the main reasons for biodiversity loss (Benton *et al.,* 2003), something that appears to be proven in the case of grasslands birds (Fuhlendorf *et al.,* 2006; Derner *et al.,* 2009). As a result, these works suggest that livestock should be used as ecosystems engineering, applying different intensities (even high grazing pressure) and combining it with other tools like fire, to maintain heterogeneous landscapes that can sustain high levels of biodiversity.

An important factor related with the homogenization of agroecosystems is the abandonment of extensive livestock management (Jongman, 2002). In many parts of the planet, especially in developed or developing countries, there is continuous depopulation of rural areas, reducing the amount of people dedicated to agriculture or farming. This is obviously reflected in the landscape, and can have noticeable negative effects in species composition or ecosystem sustainability (Torrano and Valderrábano, 2004; Altesor *et al.,* 2005; Mayer *et al.,* 2009; Farris *et al.,* 2010). Another problem in the field of biodiversity conservation, is that worldwide, grasslands are being converted to either croplands or desertified scrublands after overgrazing by livestock (Peters *et al.,* 2006).

On an oceanic island, the positive relationship between livestock management and biodiversity conservation, described by authors like Fuhlendorf et al. (2006), or Perevolotsky and Seligman, (1998), is not as clear as on the mainland. In the sensitive island's ecosystems, introduced herbivores have been considered to cause devastating effects on vegetation, and be responsible for overgrazing, ecosystem degradation, and biodiversity loss (Coblentz, 1978; Donlan *et al.,* 2002; Campbell and Donlan, 2005). Eradication of introduced herbivores has been proposed as a conservation tool in some islands. However, this proposal needs to be carefully considered, since there are complex and established ecological interactions between native and introduced species. The disappearance of these introduced species could increase damage to the ecosystems (Courchamp *et al.*, 2003). In addition, introduced herbivorous have evolved over centuries into rare native breeds, and the conservation of the valuable genetic resources of these breed varieties, at risk nowadays, has become a goal in conservation biology (Bratton, 1988; Hall and Ruane, 2003; Alderson, 2009). Moreover, we cannot negate the social, environmental and biological benefits of sustainable exploitation (Perevolotsky and Seligman, 1998), even on island environments.

The goal of this chapter is to continue the discussion of livestock management for biodiversity conservation, with special attention on the Canary Islands (Spain) as an example.

We present a review of the research done on this topic in the Canaries, looking for general conclusions and explanations of the conflicting results. Much of the research is restricted in area, however, consistent works are presented that can help us to understand the answer of the environment to grazing impact on these islands. The results of this review can provide useful information that will help managers continue with this activity, which is of high cultural value at present.

GRAZING SITUATION IN THE CANARY ISLANDS

In this archipelago, the debate about the compatibility of grazing management and biodiversity conservation becomes especially relevant. Traditional goat grazing has to cope with the environmental protection of more than the 40 % of the territory, and with more sensitive endemic species, which represent 26% of the flora (Báez *et al.,* 2001). In addition, traditional management has important socioeconomic, cultural and landscape values, recognized, for instance, by the Canarian Law of Natural Protected Spaces (Gobierno de Canarias, 1994) by means of the Rural Park figure of protection, which has the objective to conserve and promote the traditional activities (e.g. extensive grazing and agriculture), as well as the landscapes that have evolved from them (Gobierno de Canarias, 1994). Even though, it is necessary to combine both, biodiversity and traditional grazing conservation, there are few studies concerning goat management effects on the ecosystems of the Canary Islands. Furthermore, as in many other geographical areas, these studies have shown contradictory results, especially regarding the effects of grazing on species diversity and plant community composition.

Management history of the islands started with the arrival of the first inhabitants, approximately in 600 B.C. (Cabrera, 2001) and dramatically intensified after the Spanish conquest in the 15th century. In the past, land use was characterized by a transhumant grazing system, but due to the increase of tourism activity, the new protected areas created in the last decade and changes in the farming industry in the last half of the twentieth century, have led to these extensive livestock systems either being intensified or abandoned, and continuous or rotational grazing management are dominant at the present (Bermejo, 2003). Extensive grazing is carried out mainly with Canary goat breeds, evolved from the prehispanic goats into at least three genetically different native breeds (Martínez *et al.,* 2006). Traditional grazing is being subsidized by the European Common Agricultural Policy (CAP), which aims to maintain and promote traditional activities, promote a sustainable development of the outermost region, and conserve ecosystems, biodiversity and native breeds. The tendency towards the abandonment of the extensive grazing systems has not been reverted by public funding, and abandonment becomes more evident every day endangering the future of this traditional practice. At present times, majority of the herds' owners are of advanced age, and, due to the hardness of the work and the attractiveness of service sectors, this activity is not appealing to younger generations (Bermejo, 2003). Additionally, the limited seasonal pasture resources result in feeding supplementation with forages, corn, grain and fodder, implying a great dependence on foreign market products (Bermejo *et al.,* 2000). Furthermore, restrictions to open new grazing areas, due to environmental protection, make it more difficult to attract people to this activity.

GRAZING IMPACT ON VEGETATION

This has been one of the first aspects of grazing impact that have been studied in the Canary Islands. Although the published results have been only minor, a discordant discourse about the compatibility between livestock grazing and ecosystem conservation arises from them. We can classify the research done in two groups: the experimental studies, those which have been specifically designed to analyze the impact of grazing on managed ecosystems; and the observational studies, where the impact on vegetation is described, but do not contain an experimental design.

Nogales et al. (1992) studied the impact of feral goats in the native vegetation of the protected pine forest of Pajonales, Ojeada e Inagua, in the island of Gran Canaria (Figure 1). By means of direct observation of the goats and herbivore damage in the plants, they determined which plant species were consumed by goat grazing. Researchers found that a total of 24 species of phanerogams were affected by goats, 14 of these species were endemic of Gran Canaria, 4 to the Canary Island archipelago, 2 endemic to Macaronesia region and only 3 had a wider distribution. Based on these results and having taken into account that some of these species are endangered, the authors concluded that total eradication of feral goats is necessary in these protected areas. They proposed controlling goat farms around these natural spaces, setting up physical controls to avoid the entrance of animals, and the establishment of a program of genetic rescue for the endangered species.

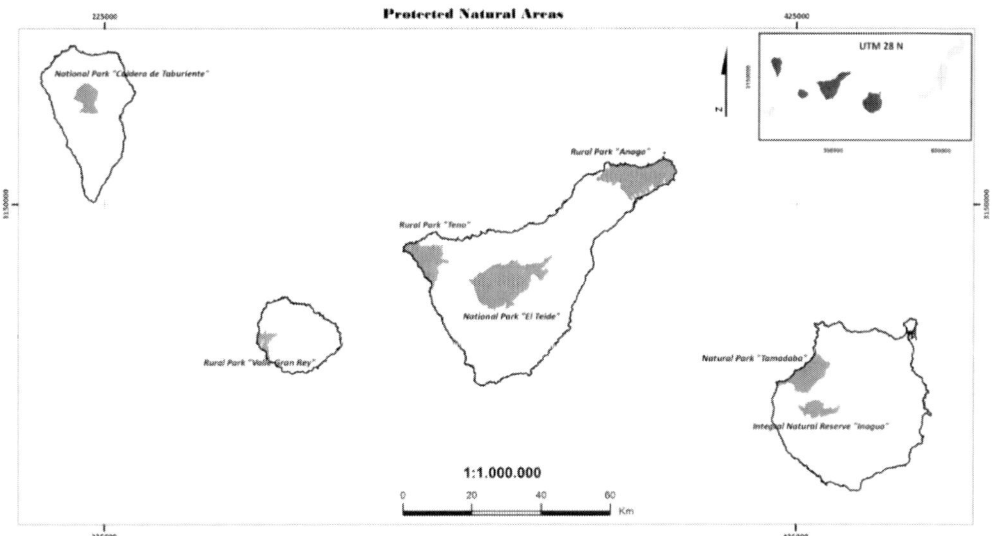

Figure 1. Canary Islands archipelago. Main areas used for grazing study are indicated.

Another observational study was carried out by Rodríguez-Piñero and Rodríguez Luengo (1993). These authors studied the impact of introduced herbivores through stomach content analyses of Corsican mouflons (*Ovis ammon musimon*) and Barbary sheep (*Ammotragus lervia*), two species introduced in the Canary Islands (in the mountain shrublands of Tenerife and the pine forest of La Palma, respectively) in the seventies for hunting aims. Nowadays, the populations of these two introduced species are mostly restricted to the Teide National Park and the surrounding Natural Park (in the case of the mouflon) in Tenerife, and in the

Caldera de Taburiente National Park in La Palma (in the case of the Barbary sheep). In addition, this study provides observational results of goats and rabbits grazing. The results of this work showed an important consumption of protected and native species by the mouflons, Barbary sheep, as well as by goats and rabbits. In spite of the few observations of the study the results were very consistent, and authors concluded that herbivores appear to be one of the primary factors negatively affecting the Canarian flora. The difficulty to control the mouflons and Barbary sheep, and their location in the National Parks, becomes a significant concern for the authors of the study. Eradication and control of these two big vertebrates is suggested in this study. In the case of these two recent introductions, eradication seems to be a reasonable conservation measure, but appears to be complicated due to the pressure of an important social sector, the hunters.

Supporting the idea of the eradication of the introduced vertebrate herbivores we would like to highlight the experimental study done by Garzón-Machado et al. (2010). These authors studied the effects of Barbary sheep, goats and rabbits on populations of four endemic legume species in pine forests in Caldera de Taburiente National Park, in the island of La Palma (Figure 1). In this research the set of endemic species was seeded along an altitudinal gradient, in plots excluded and open to grazing, and the resulting individuals were monitored over a period of four years. The result revealed a negative impact of introduced herbivores, even much stronger than expected from plant populations that have evolved without herbivore pressures, and therefore have not adaptated (Atkinson, 2001). Although this study claims the importance of small herbivores as rabbits, other studies found limited impact of herbivory in *Pinus canariensis* saplings (Arévalo *et al.,* 2009), revealing no differences in survivorship among individuals excluded from grazing and controls. The final conclusion of the research of Garzón-Machado et al. (2010) is the eradication of the Barbary sheep, goats and rabbits from the whole island, not just from the National Park. Authors state that the total eradication is the only effective long term solution to the problem. However, these proposals need to be assessed before being taken in consideration when long term introductions form part of the eradication proposal. First, there is not enough knowledge with respect to the ecological interaction established between these exotic species and native species, and the possible risk of increased damage to the ecosystems with their removal (Courchamp *et al.*, 2003). For instance, rabbits are a substantial part of the diet of two native avian scavengers, the buzzard (*Buteo buteo*) and the raven (*Corvux corax*) (Nogales and Hernández, 1994; 1997; Palacios-Palomar, 2005). Second, social implication of this measure also needs to be considered. Traditional grazing is an activity of tremendous socioeconomic and cultural value on the island of La Palma. La Palma has a genetically differentiated native goat breed (Martínez *et al.,* 2006), and the local cheese produced on this island has been officially awarded a Designation of Origin (Escuder *et al.,* 2006). Consequently, restrictions affecting this sector need to be included in the final analysis.

Gangoso et al. (2006) gather information of other observational studies in the island of Fuerteventura (Figure 1), where the number of goats have increased to more than 100,000 heads in less than 50 years (Palacios and Tella, 2003). These authors state that goat overgrazing is affecting no less than eleven endemic plant species, and that together with the droughts, typical in this dry island, overgrazing is related with increased in aridity, erosion advance and extinction of species. Nevertheless, these authors are more cautious in their conclusions, recognizing the negative effects that grazing removal can have on avian scavengers, and recommending the control of grazing exploitations, and the fencing of the most vulnerable areas.

Few studies have been developed in the islands in the last years with experimental design orientated to reveal the impact of goats in plant community species composition and species richness. These studies have been carried out mainly in traditionally managed ecosystems of several islands (Tenerife, La Gomera and La Palma), and in general they reach similar results, concluding a limited impact of traditional goat grazing in species composition as well as in species richness.

Arévalo et al. (2011a), studied the effects of grazing abandonment (more than 10 years) on species richness and species composition in pastures of the Rural Park of Valle Gran Rey, La Gomera (Figure 1). The study revealed a pattern of species composition change related to grazing, but no effects on species richness. Furthermore, there was no evidence of a relationship between the presence of exotic species and grazing management. The detected changes in species composition appear to be related to a higher abundance of native ruderal species in managed pastures. However, it is worth emphasizing that some endemic species of La Gomera (e.g. *Euphorbia bertelotii*), and endemic species of the archipelago (e.g. *Micromeria varia*, *Echium aculeatum* or *Kickxia scoparia*) were common in the grazed areas. Pasture origin in this area is related to the artificial removal of non-useful plants for agriculture more than 300 years ago (González *et al.,* 1986), and a continuous grazing management for the last 50 years. Consequently, authors expressed their concern that grazing abandonment could have a significant effect in the pastures' diversity, since grazing is a strong force in the maintenance of certain species. Due to this, they suggest promotion of this traditional activity in order to conserve natural values as well as traditional sustainable human activities.

A similar study was carried out by the same research group in the island of La Palma (Figure 1). In this case, Arévalo et al. (2011b) designed an experiment to reveal the effects of grazing on the species composition of the main vegetal communities where grazing takes place: pastures, scrubland and pine forest. This study revealed that it is necessary to integrate multiple control factors that act in each different plant community in order to be able to understand the effect of grazing (Casado *et al.,* 2004). Productivity has been considered an important variable in explaining the herbivores effects on ecosystems (Osem *et al.,* 2004; Díaz *et al.,* 2007); however the authors did not find such a relationship in these systems. Other important variables of ecosystem functioning, such as organic matter and species richness, were not significantly affected by grazing. In this case, distance to farms and altitude (as an anthropogenic and an environmental variable), were the most significant variables conditioning species composition in each different plant community, with grazing occupying a third position in affecting the studied ecosystems. Supporting the notion that specific plant communities can differently respond to grazing (Briske and Noy-Meir, 1989), in this study some vegetal communities appear to be more affected by grazing than others. Pastures showed a higher β-diversity under grazing management, this pattern, although less evident, was also found in the pastures of La Gomera (Arévalo *et al.,* 2011a). Ecosystems such as the pine forest and the scrubland were most affected by grazing, this result agrees with the common notion that scrub species become more abundant in absence of grazing (Perevolotsky and Seligman, 1999; Carmel and Kadmon, 1999; Saïd, 2001; Castro and Freitas, 2009). Although, this result appear to support those of Garzón-Machado et al. (2010) and Nogales et al. (1992), some inconsistencies can be found in terms of species response to grazing; for instance, the leguminous scrubs *Chamaecytisus proliferus* and *Adenocarpus foliolosus*, which are negatively affected by grazing in these two studies, are more common in grazed than in ungrazed areas in some of the plant communities studied by Arévalo et al. (2011b). In addition, the herbaceous legume *Lotus hillebrandii*, endemic from La Palma and characteristic of pine forest, is more abundant under grazing management. This is an example of the limitations when extrapolating the results of these types of studies among different vegetation areas (Olff and Ritchie, 1998).

In the Rural Park of Teno, in the island of Tenerife (Figure 1), different studies have been developed to analyze the impact of traditional goat grazing in species composition and diversity of managed pastures, including also the study of soil chemical composition (another aspect evaluated only in a few studies in the Canary Islands). The first study was carried out by Arévalo et al. (2007), with the main aim to determine the effects of the increase of the number of goats, experienced in the park in the nineties decade (Barquín and Chinea, 2001), on species composition, soil nutrient content and biomass nutrient content. In this study, the long term (10 years) monitoring of grazed pastures, revealed an increase in the soil phosphorous content, and a decrease of the species richness. Changes in biomass nutrient in response to grazing were not detected in the study. The increase in phosphorus was attributed to the increase in the number of goats over time. Since higher phosphorus levels have been related to the reduction of species number in pastures (Forbes *et al.,* 1980; Tracy and Sanderson, 2000), these authors concluded that it was the cause of the average species richness decrease. Due to the socioeconomic relevance of traditional grazing in the studied region, the researchers suggest promoting this activity by restoring and opening new grazing areas, in order to allow expansion in the number of animals with minimal effects on the vegetation composition and soil nutrient cycles. Due to the small scale of the sampling (species composition studied on 0.5 m x 0.5 m plots), and the fact that grazing have been proven to differently affect soil nutrients (McIntosh *et al.,* 1997; Bakker *et al.,* 2004), with other studies showing no effect or a decrease in soil phosphorus content under heavy goat grazing (Severson and Debano, 1991), the authors recognize that more studies need to be complete to determine the most adequate conservation measures.

The other two studies done in the Rural Park of Teno (Fernández-Lugo *et al.,* 2009a; 2011), used grazing exclusion as a tool to determine the possible consequences of traditional management abandonment on pasture ecosystems. After four years of goat grazing exclusion Fernández-Lugo et al. (2009a) did not find changes in species richness or diversity, but a species turnover in response to the absence of goat grazing, related to a higher soil magnesium content in the exclusion plots. The authors hypothesize that differences in magnesium between grazed and excluded plots can be related with a decrease of soil magnesium content in response to grazing due to the acceleration effects of urine deposition and soil removal on nutrient leaching. Unlike Arévalo et al. (2007) other soil nutrients, as phosphorous, were not affected by grazing exclusion, which can be explained by the fact that the increase in grazing pressure detected in the park has been reverted in the last decade (personal observation). In this study, like in that of Arévalo et al. (2011b), other factors, such as year of study and altitude, were more important in shaping plant community species composition than grazing. An interesting aspect of this research is that more exotic species were found in absence than in presence of grazing. However, it was only a pattern and not a consistent result since exotic species were poorly represented in the studied pastures. After five years of exclusion, the same authors evaluated the relationship between the effects of goat grazing exclusion on species richness and the scale of study (Fernández-Lugo *et al.,* 2011). The results revealed a higher species accumulation curves at smaller scales (0.01 to 10 m^2) in presence of grazing, indicating a positive effect of traditional management on species diversity. These studies concluded that a longer period of exclusion will be necessary to detect more relevant changes in plant species composition and that additional studies are necessary to develop a suitable pasture management strategy to maintain the stability of these ecosystems. In addition, authors suggest the promotion of goat grazing as a way to maintain land use, cultural values, and species richness in these pastures.

A couple of studies included the analysis of the productivity of the herbaceous species (Fernández-Lugo *et al.,* 2009b; de Nascimento *et al.,* 2010). The very preliminary study of Fernández-Lugo et al. (2009b), analyze the effects of goat grazing exclusion (only one year) on the species composition, species richness and aboveground primary productivity of the herbaceous species, in two Rural Parks of the islands of Tenerife and La Gomera (Figure 1). In this study, the productivity was the only variable having a significant response to grazing exclusion in such a short period of time, increasing after removal of goats. However, it was only significantly affected in the most arid pasture (La Gomera), indicating that grazing effects on productivity may depend on the precipitation range of the studied area, something that have been generally accepted (Milchunas *et al.,* 1988). On the other side, after three years of monitoring in grazed and ungrazed areas of La Palma (Figure 1), de Nascimento et al. (2010) did not find effects of traditional grazing management on net primary productivity of the herbaceous species, and neither in the species richness. The similar values of productivity in presence and absence of goats found in these studies can appear contradictory to the vegetation loss produced by goat grazing detected by Carrete et al. (2009) in Fuerteventura. Though, they can be easily explained by the most arid conditions of this eastern island, as well as by the high gradient of grazing pressures included in this study, both main factors in conditioning the response of plant communities to grazing (Milchunas and Lauenroth, 1993). Moreover, in this study the heavy grazing treatment is considerably far away from the pressure exerted by traditional extensive systems.

Although it does not focus on the impact of grazing on ecosystems, another interesting work carried out in the Canary Islands are those of Salvà-Catarineu and Romo-Díez (2008), and Otto et al. (2010), which found contradictory results with respect to the effects of goat herbivory in the regeneration of the Canarian juniper (*Juniperus turbinata canariensis*). The first study (Salvà-Catarineu and Romo-Díez, 2008), observed that herbivory by goats or sheep is an important factor affecting the regeneration of this species on the island of El Hierro. On the other hand, the second study (Otto *et al.,* 2010), did not find effects of goat pressure on tree regeneration. Another interesting review by Mata et al. (2010) propose different methodologies and analytical procedures to improve the management of goat grazed areas, recognizing that local knowledge can aid in the sustainability of goat grazing systems.

GRAZING IMPACT IN FAUNA

This subject has not been analyzed as deeply as the impact of grazing in vegetation. The only animal group in which grazing impact has been analyzed is the birds, and although the bibliography is not very extensive, several studies have revealed significant impact. Gangoso et al. (2006) did a review of the relationship between introduced herbivores and avian scavengers, and found that goat carcasses help to maintain the population of three endangered endemic subspecies of avian scavengers: the Egyptian vulture (*Neophron percnopterus majorensis*), the Eurasian buzzard (*Buteo buteo insularum*) and the common raven (*Corvus corax canariensis*). In the case of the Egyptian vulture their populations have experienced a sharp decline during the last half of the twentieth century, disappearing from the westernmost islands and surviving only on eastern islands of Fuerteventura and Lanzarote (Donázar *et al.,* 2002). The principal food sources of this species appear to be goat carcasses (Carrillo and Delgado, 1991; Medina, 1999), so it is speculated that changes in the livestock practices have had a very negative impact, being one of the reasons that have lead population of this species almost to extinction (Donázar *et al.,* 2002). The reduction of the extensive goat grazing exploitations with respect to intensive exploitations is considered responsible for the decrease of the population of this species in the islands of Lanzarote and Fuerteventura (Martín and Lorenzo, 2001). Goat carcasses also appear to be a main food source for the buzzard (Palacios-Palomar, 2005; Gangoso *et al.,* 2006), species that have suffered and extinction process parallel to the change in the farming system (Trotter, 1970). Studies on the diet of the common raven also reflect the consumption of goat carcasses (Nogales and Hernández, 1994; 1997), and the reduction of the traditional grazing systems has been also related with the regression of this species (Nogales, 1992).

After reviewing the situation of these three avian scavengers in Fuerteventura, Gangoso et al. (2006) expressed their concern about the reduction, or the removal, of the introduced herbivores as a serious threat to the raptor endemic populations of the island. The authors argue that any conservation measure implying the eradication of goats should not be implemented without a detailed knowledge of its ecological interaction with native fauna, recommending a cautious study of the communities where these non-native species can play a key role. In addition, they propose limitation of extensive grazing in areas with high number of endemic species, to decrease the effects of grazing on vegetation, and the creation of "vulture restaurants" for these scavenger birds, to decrease the dependence of population on extensive farming systems.

The short-toe larck (*Calandrella rufescens*), is another bird species which appear to be affected by grazing management, since overgrazing use to be pointed as one of the main factor affecting the habitat of this species (Viada, 1998; Illera, 2001; Illera *et al.,* 2006). Carrete et al. (2009) evaluated direct and indirect effect of goat farming on this small steppe passerine species. The results of this research showed an important number of interactions between goats and this bird species. First, heavy grazing has important effects on vegetation cover loss, determining habitat impoverishment for this species. Second, this negative impact is, in part, compensated by the food supply that this species can find in the farms. Third, beside of the positive effects that the availability of a constant food resource can have on the populations, the change of the food behavior of this bird appears to be related to an increase in the nutritious and infectious diseases. The authors concluded that urgent measures are necessaries to guarantee the conservation of the species. They propose fencing farms to avoid short-toe larks feeding in them, limiting goat densities to avoid overgrazing, and restore the habitat of the species. They emphasize that these measures should be carried out together to avoid negative effects of a decreasing food supply in the populations. Additionally, authors recognize that eradication of goat from this island is unfeasible due to the socioeconomic and cultural implications.

CONCLUSION

Throughout this chapter we have demonstrated some of the conflicting results on the impact of grazing on Canarian ecosystems. But, analyzing the presented works we can find some explanations to these contradictions. First, the existence of two kinds of works, those carried out in "pristine" ecosystems, and those done in traditionally managed ecosystems. The first ones use to have high degree of endemism and population of endangered species. Endemic species of the Canarian flora have evolved without herbivore pressures, and maybe have not adaptations to it (Atkinson, 2001), explaining their higher sensitivity to goat grazing. Moreover, endangered species, in general, have suffered the reduction of their populations due to a combination of anthropogenetic factors (removal of woody species, fragmentation of ecosystems, fire, agriculture and farming), being in a very sensitive state, in which any damage, as herbivory, can be clearly negative. On the other hand, traditionally managed ecosystems are dominated by native herbaceous and shrub species, mainly of a Mediterranean origin (Santos, 2001; del Arco *et al.,* 2010), that can be more resilient to grazing pressure (Lavorel, 1999). Second, the broad range of climatic conditions existing on the Archipelago. These conditions varied from arid to humid, determining the existence of different types of vegetal communities. Both factors, specific plant community (Briske and Noy-Meir, 1989) and environmental gradient (Milchunas *et al.,* 1988; Osem *et al.,* 2002), condition the response of the vegetation to grazing. Third, the extensive grazing systems, the intensive grazing systems, and the feral goats, (differently represented in the reviewed works), can exert varying degrees of pressure on the territory (Bermejo, 2003). This becomes a fundamental factor in determining grazing effects on ecosystems (Milchunas and Lauenroth, 1993). Finally, in the case of the fauna, the difference between positive and negative impact of grazing in bird populations can be easily explained by differences in the group of birds studied.

Although some negative impacts of grazing have been revealed, in general they have been restricted to some specific areas or species. In those cases where goat grazing is threatening endemic and endangered populations, control of goat densities, grazing restrictions and use of fences should be encourage, however drastic measures as eradication of goats are not ecologically or socially feasible. On the other hand, grazing of small goat herds on traditionally managed ecosystems seems to be a sustainable activity, which removal have been related with a reduction in the number of species, changes in species composition and also an increase in exotic species.

The government of The Canary Islands considers goat grazing as both a traditional and necessary activity as the local economy depends on it, but also the derived products such as the goat cheese with Origin Designation, local varieties of fodder and native goat breeds are completely related and dependent from the maintenance of this activity. For this reason, and in order to avoid the disappearance of this activity, we suggest the promotion of goat grazing in traditionally managed ecosystems, applying a strict control of grazing pressure and type of management, as a way of maintaining landscape use, cultural values and biodiversity under typical grazing pressure.

ACKNOWLEDGMENTS

Authors thank financial and logistic support by the Cabildo de La Gomera, Cabildo de La Palma, Cabildo de Tenerife and Cabildo de Gran Canaria. Also, the University of La Laguna (through the ULL-Research Support Program) and the Spanish Ministry of Science and Innovation (project CGL2006-06471 and FPU scholarship for Silvia Fernández Lugo) offered partial financial support. We are grateful to Javier Mata, and the people who helped with our fieldwork. We specially would like to thank goat keepers of the islands for their cooperation with the projects of the group.

REFERENCES

Altesor, A., Oesterheld, M., Leoni, E., Lezama, F., and Rodríguez, C. (2005) Effect of grazing exclosure on community structure and productivity of a Uruguayan grassland. *Plant Ecology, 179,* 83-91.

Alderson, L. (2008) Breeds at risk: Definition and measurement of the factors which determine endangerment. *Livestock Science, 16,*1170-1175.

Atkinson, I.A.E. (2001) Introduced mammals and modesl for restoration. *Biological Conservation, 99,* 81-96.

Arévalo, J.R., Chinea, E., and Barquín, E. (2007) Pasture management under goat grazing in Canary Islands. *Agriculture, Ecosystems and Environment, 118,* 291-296.

Arévalo, J.R. (2009) Treatments on the survival of *Pinus canariensis* Chr. Sm. Ex DC in Buch planted seedlings in arid zones (herbivores protector-fertilizer-mulch-hydrogels). *The Open Forest Science Journal, 2,* 25-30.

Arévalo, J.R., de Nascimento, L., Fernández-Lugo, S., Saro, I., Camacho, A., Mata, J., and Bermejo, L. (2011a) Effects of abandoning long-term goat grazing on species composition and species richness of pastures at La Gomera, Canary Islands. *Spanish Journal of Agriculture, 9,* 113-123.

Arévalo, J.R., de Nascimento, L., Fernández-Lugo, S., Mata, J., and Bermejo, L.A. (2011b) Grazing effects on species composition in different plant communities (La Palma, Canary Islands) *Acta Oecologica, 37,* 230-238.

Aviron, S., Nitsch, H., Jeanneret, P., Buholzer, S., Luka, H., Pfiffner, L., Pozzi, S., Schupbach, B., Walter, T., and Herzog, F. (2009) Ecological cross compliance promotes farmland biodiversity in Switzerland. *Frontiers in Ecology and the Environment, 7,* 247-252.

Báez, M., Martín-Esquivel, J.L., and Oromí, P. (2001) Diversidad taxonómica terrestre. In J.M, Fernández-Palacios, and J.L. Martín Esquivel (Eds.*), Naturaleza de las Islas Canarias, Ecología y Conservación* (pp. 119-125) Santa Cruz de Tenerife, Spain: Editorial Turquesa.

Bakker, E.S., Olff, H., Boekhoff, M., Gleichman, J.M., and Berendse, F. (2004) Impact of herbivores on nitrogen cycling: contrasting effects of small and large species. *Oecologia, 138,* 91-101.

Barquín, E., and Chinea, E. (1991) The Teno plateau, Tenerife (Canary Islands), a study of this singular ecosystem, traditionally managed. In *Proceedings of the XXXI reunión científica de la SEEP* (pp. 384-388) Murcia, Spain.

Baydack, R.K., and Campa, III H.C. (1999) Setting the context. In R.K. Baydack, III H.C., Campa, and J.B. Haufler (Eds.), *Practical Approaches to the Conservation of Biological Diversity* (pp. 3-16). Washington, D.C.: Island Press.

Benton, T.G., Vickery, J.A., and Wilson, J.D. (2003) Farmland biodiversity: is habitat heterogeneity the key? *Trends in Ecology and Evolution, 18,* 182-188.

Bermejo, L.A. (2003) Conservación de los recursos genéticos caprinos en los espacios protegidos de Canarias: Impacto social y ambiental. (PhD thesis), Universidad de Córdoba, Spain.

Bermejo, L.A., Mata, J., Delgado, J.V., Flores, M.P., and Camacho, A. (2000) Uso ganadero del Parque Rural de Anaga: resultados preliminares. *Archivos de Zootecnia, 49,* 269-274.

Bratton, S.P. (1988) Minor Breeds and Major Genetic Losses. *Conservation Biology, 2,* 297-299.

Brinkmann, K., Patzelt, A., Dickhoefer, U., Schlecht, E., and Buerkert, A. (2009) Vegetation patterns and diversity along an altitudinal and a grazing gradient in the Jabal al Akhdar mountain range of northern Oman. *Journal of Arid Environments, 73,* 1035-1045.

Briske, D.D., and Noy-Meir, I. (1998) Plant response to grazing: a comparative evaluation of annual and perennial grasses. In V.P. Papanastasis, and D. Peter (Eds.), *Ecological Basis of Livestock Grazing in Mediterranean Ecosystems* (pp. 13-26). Luxembourg: European Commision, Science, Research and Development.

Cabrera, J.C. (2001) Poblamiento e impacto aborigen. In J.M, Fernández-Palacios, and J.L. Martín Esquivel (Eds.*), Naturaleza de las Islas Canarias, Ecología y Conservación* (pp. 241-245). Santa Cruz de Tenerife, Spain: Editorial Turquesa.

Campbell, K., and Donlan, C.J. (2005) Feral goat eradications on islands. *Conservation Biology, 19,* 1362-1374.

Carmel, Y., and Kadmon, R. (1999) Effects of grazing and topography on long-term vegetation changes in a Mediterranean ecosystem in Israel. *Plant Ecology, 145,* 243-254.

Carrillo, J., and Delgado, G. (1991) Threats to and conservationist aspects of birds of prey in the Canary Islands. *Birds of Prey Blulletin, 4,* 25-32.

Carrete, M., Serrano, D., Illera, J.C., López, G., Vögeli, M., Delgado, A., and Tella, J. (2009) Goats, birds, and emergent diseases: apparent and hidden effects of exotic species in an island environment. *Ecological Applications, 19,* 840-853.

Castro, H., and Freitas, H. (2009) Aboveground biomass and productivity in the Montado: from herbaceous to shrub dominated communities. *Journal of Arid Enviroments, 73,* 506-511.

Casado, M.A., Castro, I., Ramírez-Sanz, L., Costa-Tenorio, M., de Miguel, J.M., and Pineda, F.D. (2004) Herbaceous plant richness and vegetation cover in Mediterranean grasslands and shrublands. *Plant Ecology, 170,* 83-91.

Clergue, B., Amiaud, B., Pervanchon, F., Lasserre-Joulin, F., and Plantureux, S. (2005) Biodiversity: function and assessment in agricultural areas. A review. *Agronomy for Sustainable Development, 25,* 1-15.

Coblentz, B.E. (1978) The effect of feral goats (Capra hircus) on island ecosystems. *Conservation Biology, 4,*261-265.

Courchamp, F., Chapuis, J.L., and Pascal, M. (2003) Mammal invaders on islands: impact, control and control impact. *Biological Reviews, 78,* 347-383.

De Bello, F., Lepš, J., and Sebastià, M.T. (2007) Grazing effects on species-area relationship: variation along a climatic gradient in NE Spain. *Journal of Vegetation Science, 18,* 25-34.

De Nascimento, L., Bermejo, L.A., Saro, I., Fernández-Lugo, S., and Mata J. (2010) Relationship between herbaceous productivity and species richness in grazed areas on La Palma (Canary Islands). In H. Schnyder, J. Isselstein, F. Taube, K. Auerswald, J. Schellberg, M. Wachendorf, A. Herrmann, M. Gierus, N. Wrage, A. Hopkins (Eds.), *Grassland in a changing world. Vol 15: Grassland science in Europe* (pp 925-927). Kiel, Germany: European Grassland Federation.

Del Arco, M., González-González, R., Garzón-Machado, V., and Pizarro-Hernández, B. (2010) Actual and potential natural vegetation on the Canary Islands and its conservation status. *Biodiversity and Conservation, 19,* 3089-3140.

Derner, J.D., Lauenroth, W.K., Stapp, P., and Augustine, D.J. (2009) Livestock as ecosystems engineers in semiarid rangelands: Addressing grasslands birds habitat in the Western Great Plains of North America. *Rangeland Ecology and Management, 62,* 111-118.

Díaz, S., Lavorel, S., McIntyre, S., Falczuk, V., Casanoves, F., Milchunas, D. G., Skarpe, Ch., Sternberg, G.R., Noy-Meir, I., Landsberg, J., Zhang, W., Clark, H., and Campbell, B. (2007) Plant traits responses to grazing: a global synthesis. *Global Change Biology, 13,* 313-341.

Donázar, J.A., Palacios, C.J., Gangoso, L., Ceballos, O., González, M.J., and Hiraldo, F. (2002) Conservation status and limiting factors of the endangered population of Egyptian Vulture (*Neophron percnopterus*) in the Canary Islands. *Biological Conservation, 107,* 89-98.

Donlan, C.J., Tershy, B.R., and Croll, D.A. (2002) Islands and introduced herbivores: conservation action as ecosystem experimentation. *Journal of Applied Ecology 39,* 235-246.

Eken, G., Bennun, L., Brooks, T.M., Darwall, W., Fishpool, L.D.C., Foster, M., Knox, D., Langhammer, P., Matiku, P., Radford, E., Salaman, P., Sechrest, W., Smith, M.L., Spector, S., and Tordoff, A. (2004) Key biodiversity areas as site conservation targets. *Bioscience, 54,* 1110-1118.

Escuder, A., Fernández, G., and Capote, J. (2006) Characterization of Palmera dairy goats production systems. *Options Méditerranéennes, 70,* 95-100.

Farris, E., Filibeck, G., Marignani, M., and Rosati, L. (2010) The power of potential natural vegetation (and of spatial–temporal scale): a response to Carrión and Fernández (2009) *Journal of Biogeography, 37,* 2211-2213.

Fernández-Lugo, S., de Nascimento, L., Mellado, M., Bermejo, L.A, and Arévalo, J.R. (2009a) Vegetation change and chemical soil composition after four years of goat grazing exclusion in a Canary Islands pasture. *Agriculture, Ecosystems and Environment, 132,* 276-282.

Fernández-Lugo, S., de Nascimento, L., Saro, I., Bermejo, L.A., Arévalo, J.R. (2009b) Efectos de la exclusión del pastoreo en la diversidad, riqueza, composición de especies y productividad de dos pastizales canarios. In R. Reiné, O. Barrantes, A. Broca, and C. Ferrer (Eds.), *La multifuncinalidad de los pastos: producción ganadera sostenible y gestión de los ecosistemas* (pp. 155-161). Huesca, Spain: S.E.E.P.

Fernández-Lugo, S., de Nascimento, L., Mellado, M., and Arévalo, J.R. (2011) Grazing effects on species richness depends on scale: a 5 year study in Tenerife pastures (Canary Islands). *Plant Ecology, 212*, 423-432.

Forbes, T.J., Dibb, C., Green, J.O., Hopkings, A., and Peel, S. (1980) *Factors affecting the productivity of permanent grassland; a national farm study*. Joint permanent pasture group, Hurley, Maidenhead, Berks.

Fraser, I., and Chisholm, T. (2000) Conservation or cultural heritage? Cattle grazing in the Victoria Alpine National Park. *Ecological Economics, 33*, 63-75.

Fuhlendorf, S. D., and Smeins, F. E. (1999) Scaling effects of grazing in a semiarid grassland. *Journal of Vegetation Science, 10*, 731-738.

Fuhlendorf, S.D., Harrell, W.C., Engle, D.M., Hamilton, R.G., Davis, C.A., and Leslie, D.M. (2006) Should heterogeneity be the basis for conservation? Grassland bird response to fire and grazing. *Ecological Applications, 16*, 1706-1716.

Gangoso, L., Donázar, J., Scholz, S., Palacios, C.J., and Hiraldo, F. (2006) Contradiction in conservation of island ecosystems: Plants, introduced herbivores and avian scavengers in the Canary Islands. *Biodiversity and Conservation, 15*, 2231- 2248.

Garzón-Machado, V., González-Mancebo, J.M., Palomares-Martínez, A., Acevedo-Rodríguez, A., Fernández-Palacios, J.M., del-Arco-Aguilar, M., Pérez de Paz, P.L. (2010) Strong negative effect of alien herbivores on endemic legumes of the Canary pine forest. *Biological Conservation, 143*, 2685-2694.

Gobierno de Canarias (1994) Ley 12/1994, de 19 de diciembre, de Espacios Naturales de Canarias. Boletín Oficial de Canarias ,24 de Diciembre de 1994, Nº157.

González, M.N., Rodrigo, J.D., and Suárez C. (1986) *Flora y vegetación del archipiélago canario*. Las Palmas de Gran Canaria, Spain: Edirca S.L.

Hall, S.J.G., and Ruane, J. (1993) Livestock Breeds and Their Conservation: A Global Overview. *Conservation Biology, 7*, 815-825.

Hill, J., Hostert, P., Tsiourlis, G., Kasapidis, P., and Udelhoven, Th. (1998) Monitoring 20 years of intense grazing impact on the Greek island of Crete with earth observation satellites. *Journal of Arid Environments, 39*, 165-178.

Illera, J.C. (2001) Habitat selection by the Canary Islands stonechat (Saxicola dacotiae) (Meade-Waldo, 1889) in Fuerteventura Island: a two-tier habitat approach with implications for its conservation. *Biological Conservation*, 97, 339-345.

Illera, J.C., Díaz, M., and Nogales, M. (2006) Ecological traits influence the current distribution and range of an island endemic bird. *Journal of Biogeography, 33*, 1192-1201.

Jongman, R.H.G. (2002) Homogenisation and fragmentation of the European landscape: ecological consequences and solutions. *Landscape and Urban Planning, 58*, 211- 221.

Krebs, J.R., Wilson, J.D., Bradbury, R.B., and Siriwardena, G.M. (1999) The second Silent Spring? *Nature, 400*, 611-612.

Lavorel, S. (1999) Ecological diversity and resilience of Mediterranean vegetation to disturbance. *Diversity and Distributions, 5*, 3-13.

McIntosh, P.D., Allen, R.B., and Scott, N. (1997) Effects of exclosure management on biomass and soil nutrient pools in seasonally dry high country, New Zealand. *Journal of Environmental Management, 52*, 169-186.

Martín, A., and Lorenzo, J.A. (2001) *Aves del Archipiélago Canario*. San Cristobal de La Laguna, Spain: F. Lemus Editorial.

Martínez, A., Acosta, J., Vega-Pla, J.L., and Delgado, J.V. (2006) Analysis of the genetic structure of the Canary goat populations using microsatellites. *Livestock Science, 102,* 100-140.

Mata, J., Bermejo, L.A., de Nascimento, L., and Camacho, A. (2010) The problem of grazing planning in a non-equilibrated environment, from the analytical procedure toward the system approach. *Small Ruminant Research, 89,* 91-101.

Mayer, J.J., and Brisbin, Jr. I.L. (Editors) (2009) *Wild pigs: Biology, damage, control techniques and management–SRNL-RP-2009-00869.* Aiken, South Carolina: Savannah River National Laboratory.

Medina, F.M. (1999) Alimentación del alimoche, *Neophron percnopterus* (L.), en Fuerteventura, Islas Canarias (Aves, Accipitridae). *Vieraea, 27,* 77-86.

Milchunas, D.G., Sala, O.E., and Lauenroth, W.K. (1988) A generalized model of the effects of grazing by large herbivores on grasslands community structure. *The American Naturalist, 132.* 87-106.

Milchunas, D.G., and Lauenroth, W.K. (1993) Quantitative effects of grazing on vegetation and soil over a global range of environments. *Ecological Monographs, 63,* 327-366.

Millennium Ecosystem Assessment (2005) *Ecosystems and Human Well-being: Biodiversity Synthesis.* Washington, DC: World Resources Institute.

Nogales, M. (1992) Problemática conservacionista del cuervo *(Corvus corax)* en Canarias y estado de sus distintas poblaciones. *Ecología, 6,* 215-223.

Nogales, M., Marrero, M., and Hernández, E.C. (1992) Efectos de las cabras cimarronas (*Capra hircus L.*) en la flora endémica de los pinares de Pajonales, Ojeda e Inagua (Gran Canaria). *Botánica Macaronésica, 19-20,* 79-86.

Nogales, M., and Hernández, E. C. (1994) Interinsular variations in the spring and summer diet of the raven *Corvus corax* in the Canary Islands. *Ibis, 136,* 441-447.

Nogales, M., and Hernández, E.C. (1997) Diet of common ravens on El Hierro, Canary Islands. *Journal of Field Ornithology, 68,* 382-391.

Olff, H., and Ritchie, M.E. (1998) Effects of herbivores on grassland plant diversity. *Trends in Ecology and Evolution, 13,* 261-265.

Osem, Y., Perevolotsky, A., and Kigel, J. (2002) Grazing effect on diversity of annual plant communities in a semi-arid rangeland: interactions with small-scale spatial and temporal variation in primary productivity. *Journal of Ecology, 90,* 936-946.

Osem, Y., Perevolotsky, A. and Kigel, J. (2004) Site productivity and plant size explain the response of annual species to grazing exclusion in a Mediterranean semi-arid rangeland. *Journal of Ecology, 92,* 297-309.

Otto, R., Krüsi, B., Delgado, J.D., Fernández-Palacios, J.M., Garcia-del-Rey, E., and Arevalo, J.R. (2010) Regeneration niche of the Canarian juniper: the role of adults, shrubs and environmental conditions. *Annals of Forest Science,* DOI: 10.105/forest/2010034.

Palacios, C.J., and Tella, J.L. (2003) Aves esteparias de Canarias: una comunidad en peligro. *Quercus, 209,* 14-21.

Palacios-Palomar, C.J. (2005) El ratonero común (*Buteo buteo insularum*) en Fuerteventura, Islas Canarias (Aves, Accipitridae). *Vieraea, 33,* 1-5.

Peco, B., Sánchez, A.M., and Azcárate, F.M. (2006) Abandonment in grazing systems: Consequences for vegetation and soil. *Agriculture, Ecosystems and Environment, 113,* 284-294.

Perevolotsky, A., and Seligman, N.G. (1998) Role of grazing in Mediterranean rangeland ecosystems. Inversion of a paradigm. *Bioscience, 48,* 1007-1017.

Peters, D.P.C., Bestelmeyer, B.T., Herrick, J.E., Monger, H.C., Fredrickson, E., and Havstad, K.M. (2006) Disentangling complex landscapes: new insights to forecasting arid and semiarid system dynamics. *BioScience, 56,* 491-501.

Rodríguez-Piñero, J.C., and Rodríguez-Luengo, J.L. (1993) The effect of herbivores on the endemic Canary flora. *Boletin do Museo Municipal do Funchal, 2,* 265-271.

Saïd, S. (2001) Floristic and life form diversity in post-pasture successions on a Mediterranean island (Corsica). *Plant Ecology, 162,* 67-76.

Salvà-Catarineu, M., and Romo-Díez, A. (2008) Uso de TIG para la conservación del sabinar de *Juniperus turbinata* subsp. *canariensis* en la isla de El Hierro. In L., Hernández, and J.M., Parreño (Eds.), *Tecnologías de la Información Geográfica para el Desarrollo Territorial* (pp. 766-776.). Las Palmas de Gran Canaria, Spain: Servicio de Publicaciones y Difusión Científica de la ULPGC.

Santos, A. (2001) Flora vascular. In: J.M. Fernández-Palacios, and J.L. Martín (Eds.), *Naturaleza de las Islas Canarias. Ecología y Conservación* (pp. 187-192). S/C de Tenerife, Spain. Publicaciones Turquesa.

Sanderson, M.A., Skinner, R.H., Barker, D.J., Edwards, G.R., Tracy, B.F., and Wedin, D.A. (2004) Plant species diversity and management of temperate forage and grazing land ecosystems. *Crop Science, 44,* 1132-1144.

Severson, K.E., and Debano, L.F. (1991) Influence of Spanish goats on vegetation and soils on Arizona chaparral. *Journal of Range Management, 44,* 111-117.

Stohlgren, T.J., Schell, L.D., and Vanden Heuvel B. (1999) How grazing and soil quality affect native and exotic plant diversity in rocky mountain grasslands. *Ecological Applications, 9,* 45-64.

Swagemakers, P., Wiskerke, H., and Van Der Ploeg, J.D. (2009) Linking birds, fields and farmers. *Journal of environmental management, 90,* 185-192.

Torrano, L., and Valderrábano, J. (2004) Review. Impact of grazing on plant communities in forestry areas. *Spanish Journal of Agricultural Research, 2,* 93-105.

Tracy, B.F., and Sanderson, M.A. (2000) Patterns of plant species richness in pasture lands of the northeast United States. *Plant Ecology, 149,*169-180.

Trotter, W.D.C. (1970) Observations faunistiques sur lille de Lanzarote (Canaries). *L'Oiseau et le RFO, 40,* 160-172.

Viada, C. (1998) *Áreas Importantes para las Aves en España* (2nd edition). Madrid, Spain: SEO/ Birdlife.

Wilcove, D. (1993) Getting ahead of the extinction curve. *Ecological Applications, 3,* 218-220.

Zhang, J.T., and Dong, Y. (2010) Factors affecting species diversity of plant communities and the restoration process in the loess area of China. *Ecological Engineering, 36,* 345-350.

In: Grazing Ecology: Vegetation and Soil Impact
Editor: José Ramón Arévalo
ISBN 978-1-62100-436-3

Chapter 5

TEN YEARS OF EXCLUSION FROM GRAZING INCREASE NATIVE PLANT DIVERSITY IN THE AUSTRAL PATAGONIAN STEPPE

***Erwin Domínguez*[1] *and Aníbal Pauchard*[2*]**

[1]Instituto de Investigaciones Agropecuarias (INIA – CRI Kampenaike), Casilla 277, Punta Arenas, Chile and Centro de Estudios del Cuaternario de Fuego - Patagonia y Antártica (CEQUA)

[2]Facultad de Ciencias Forestales, Universidad de Concepción and Instituto de Ecología y Biodiversidad (IEB). Casilla 160-C, Concepción, Chile

ABSTRACT

Grazing and its interaction with other environmental variables generate temporal and spatial variations in plant communities. In the Patagonian steppe in southern Chile, two conditions were analyzed: a grazed active cattle ranch and a non-grazed area protected from domestic cattle during a 10-year period. The aim was to determine how the exclusion from domestic grazing affects species diversity and the replacement grasses of *Festuca gracillima*, with dwarf shrublands of *Empetrum rubrum*. Species richness, foliage coverage, and percent of bare soil at $1 m^2$ and $1000\ m^2$ scales were compared using MANOVA, to test possible differences due to grazing effects and edaphic conditions. Multivariate analyses (Jaccard's similarity coefficients and Detrended Correspondence Analysis) were used to test for community differences.

The results show that the *F. gracillima* grasses had higher species richness in non-grazed sites compared to the shrublands. At the $1000\ m^2$ scale, no significant difference in exotic species richness was found between the non-grazed and the grazed sites. Soil fertility was significantly higher in grasses with respect to shrublands at the $1000\ m^2$ scale. Jaccard's Index indicated a 49 % of floristic similarity between grazed and non-grazed grasses at the $1000\ m^2$ scale. The DCA at the $1\ m^2$ scale with presence/absence data indicated a continuous gradient with respect to the floristic composition between grass and shrublands species, independent of the grazing factor. A similar response was observed at the $1000\ m^2$ scale.

[*] Corresponding author Email: pauchard@udec.cl

This research concludes that protection from grazing increased native species richness at the 1 and 1000 m^2 scales in only one decade. These rapid changes are most probably caused by the release from excessive grazing, which is threatening the conservation of plant species in the Austral Patagonian steppe.

Keywords: Natural prairies degradation, inadequate grazing rate, species diversity conservation

Abbreviations

Gr, grazed;
NGr, non-grazed;
S, dwarf shurblands;
G, grassland;
AE, dwarf shrublands (prostrated or cushioned);
AM, medium shrublands (straight);
GA, annual grasses (Poaceae);
GP, perennial grasses (Poaceae);
GM, graminoids (cyperaceous and juncaceous);
HA, annual herbs;
HP, perennial herbs.

Introduction

The effect of grazing on plant community diversity has been widely studied in terrestrial ecosystems. The increase or reduction in grazing is the principal disturbance in natural prairies (Milchunas *et al.,* 1988; Milchunas and Lauenroth, 1993; Proulx and Mazumber, 1998; Stohlgren *et al.,* 1999[a]), producing important changes in the diversity of native and exotic species in functional plant groups at the community and ecosystem level (Bakker 1989, 1999; Belsky, 1992; Breeskow *et al.,* 1995; Proulx and Manzumbre, 1998; Davis *et al.,* 2000; Adler *et al.,* 2005). At moderate grazing levels, greater species diversity is expected according to the hypothesis of intermediate disturbance (Grime, 1973; Proulx and Manzumbre, 1998). However, this does not always occur because the increased diversity will generally depend on resource availability and the evolutionary grazing history, which differs for each particular prairie ecosystem (Milchunas *et al.,* 1988; Milchunas and Lauenronth, 1993; Olff and Ritchie, 1998; Díaz *et al.,* 1994; Cingolani *et al.,* 2003; Adler *et al.,* 2005).

The grazing effect and its interaction with other environmental variables generate temporal and spatial variations in species richness in plant communities (Osem *et al.,* 2002; Stohlgren *et al.,* 1999[a]; Milchunas *et al.,* 1988; Tilman, 1988; Huisman *et al.,* 1999; Collantes *et al.,* 1999; Stohlgren *et al.,* 1999[b]; Cingolani *et al.,* 2003, 2005). This result is fundamentally due to the spatial heterogeneity in the distribution of environmental factors such as the availability of nutrients, precipitation, light and topographical characteristics among other (Mackey and Curie, 2001; Safford and Harrison, 2001).

Grazing causes structural changes in the plant community through the modification of the growth forms of the dominant species and their foliage coverage (Pykälä, 2004), which differs at the spatial scale depending on the individual species' response (Stohlgren *et al.*, 1999[a], 1999[b]; Posse *et al.*, 2000; Saffor and Harri, 2001). Many of these changes are related to the replacement of grasses by shrublands because grasses are preferred by the herbivores and because steppe shrublands are adapted to survive in arid environments characterized by low resource availability (Milchunas *et al.*,1988; Lauenroth and Aguilera, 1998; Hulme *et al.*, 1999; Watkinson and Ormedod, 2001; Landsberg *et al.*, 2003; Cingolani *et al.*, 2005). This change produces a serious problem for arid and semi-arid prairie ecosystem conservation worldwide. Only in South Africa, 13,000 hectares have been invaded by thorny bushes (Roques *et al.*, 2001). A similar situation has been reported for other ecosystems (Scholes and Archer, 1997). Such a response has been found in South America in the central region of Argentina, where grazing has reduced grass coverage, promoting the establishment of shrublands (Bisigoto and Bertiller, 1997, 2004).

For the steppes located in northern Tierra del Fuego in Argentina, it is recognized that the over-grazing at the beginning of the 20th century could be the direct cause of the replacement of natural grasses by dwarf shrublands (Baetti *et al.*, 1993; Mendoza *et al.*, 1995). This replacement occurred especially in soils of glacier origin (Pleistocene), which are poor in nutrients and acids (Collantes et al. 1989). In this area, similar to the European heathland in their physiognomy and edaphic characteristics, the dwarf shrublands of *Empetrum rubrum* have become important (Collantes *et al.*, 1989; Posse *et al.*, 1996; Collantes *et al.*, 1999). Additionally, shrublands are characterized by their low productivity and diversity (Collantes *et al.*, 1989; Posse *et al.*, 1996; Cingolani *et al.*, 1998; Collantes *et al.*, 1999; Posse *et al.*, 2000). The evidence suggests that an important surface area of northern Tierra del Fuego presently with grasses is at risk to be replaced by dwarf shrublands (Borrelli and Oliva, 2001), which could cause important changes in native and exotic plant diversity if the actual grazing intensity is maintained.

To characterize the grazing effect on the diversity of dwarf shrublands *Empetrum rubrum* and grasses of *Festuca gracillima* located on the continental portion of the eastern branch of the Magellan's Strait, our study was carried out in the only protected area where no domestic grazing has been permitted in the past 10 years. This protected area is the largest of its type in the southern Chile Patagonian steppe (Domínguez *et al.*, 2004) and is surrounded by cattle ranches where the two studied communities are found, growing in similar environmental conditions (e.g., topographic, pedogenic, and climatic). They have a 120-year domestic grazing history, sufficient to produce irreversible changes in the vegetation communities (Cingolani *et al.*, 2005). Consequently, it is difficult to determine if the presently dominant vegetal communities are the product of over-grazing (e.g., dwarf shrublands) or the product of an environmental type different from the grasses (Collantes *et al.*, 1989; Collantes *et al.*, 1999).

We address the following questions: 1) How do the exclusion from domestic grazing affect species diversity and promote the replacement of grasses such as *Festuca gracillima*, with dwarf shurblands of *Empetrum rubrum*? We intend to evaluate how the exclusion from domestic grazing affects the native and exotic species richness and foliage coverage using a nested design of standardized sampling at a multiple spatial scale, 2) How do floristic similarity between dwarf shrublands and grasses change between grazed and non-grazed sites, and 3) is there physiochemical differences in the soil between shrublands and grasses under grazed and non-grazed sites?

METHODS

Study Area

The study was carried in the Pali Aike National Park and on the Bernardo O'Higgins cattle cooperative that surrounds the park (52° 04' y 52° 05' S and 69° 47' y 69° 51' W - 110 a 250 m) in XII Region, Chile (Figure 1). The Pali Aike National Park was created to conserve a representative portion of the Patagonian steppe ecosystem from the domestic grazing effect by fencing in its 5,030 hectares. The climate corresponds to Cold Steppe with mean temperatures between 3° and 4° C, characterized by strong winds and mean annual precipitation between 200 and 250 mm, distributed heterogeneously throughout the year (Endlicher and Santana, 1988). The water balance has a notable deficit in December and January due to the high velocity winds, which produce a high potential rate of evapotranspiration (Koremblit and Force Ley, 1991; Walter and Box, 1983). The soil type is Prairie -Planosol (Sáez, 1995). The topography is integrated by planes with soft waves that have been modeled by glacial action and fluvioglacial erosion. Its ecosystem corresponds to the biotic province of the Austral Patagonian Steppe (Pisano, 1977), the dominant vegetal communities are natural grasses of *Festuca gracillima* and dwarf shrublands of *Empetrum rubrum.* Park vegetation supports the grazing of wild animals such as: *Lama guanicoe* (guanacos), *Pterocnemia pennata* (ñandúes), and *Chloephaga picta* (caiquenes). On the ranch, sheep raising (*Corriedale race*) has been an important activity for 120 years.

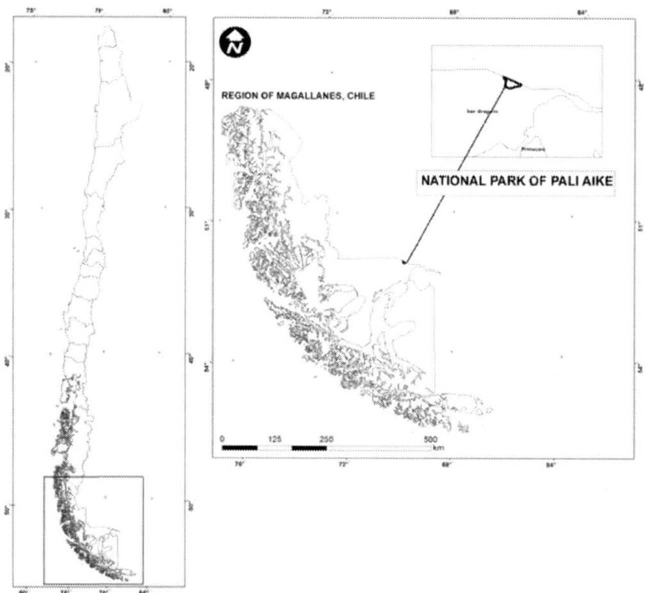

Figure 1. Location of the study area in the XII Region, Chile.

Methods

Sampling was conducted in two contrasting areas. The first is an area with 10 years of effective exclusion from domestic grazing inside the Pali Aike National Park. The second is a situation of extensive grazing with an intensity of 0.76 AU/ha with a slow rotation (SAG 2004). In both treatments, eight 1500 x 1000 m (150 hectares) blocks were established (4 in grasses and 4 in shrublands).Within each block 6 1000 m^2 plots (0.1-hectare) were randomly located: 24 plots inside the Pali Aike National Park and 24 outside the Park, located in the C.B. O'Higgins cattle ranch. The vegetation sampling followed a modified nested Whittaker design (Stohlgren *et al.,* 1998). The plots were geo-referenced using a global positioning system (GPS, Garmin Emap and the software Oziexplorer).

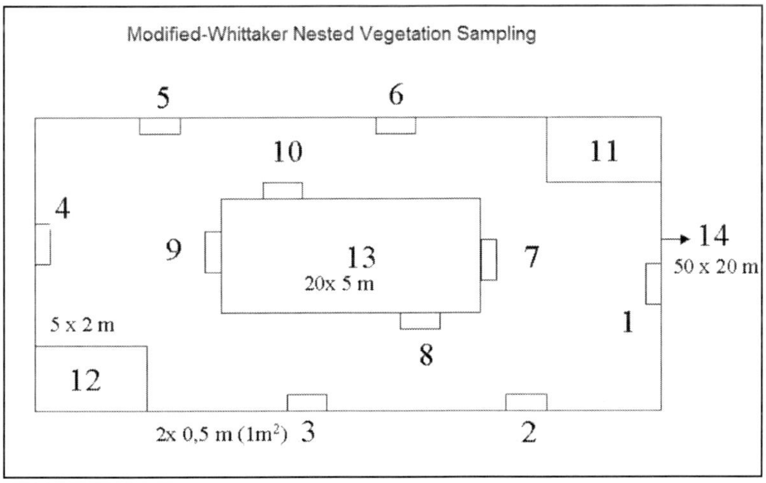

Figure 2. Schematic diagram of the nested modified Whittaker plots for vegetation sampling (Stohlgren et al. 1998).

To measure the plant community patterns, Whittaker modified plots were established (Figure 2, Stohlgren *et al.,* 1998). This design implies a 1000-m^2 plot with nested plots that consisted in one 100 m^2, two 10 m^2 plots and 10 sub-plots of 1 m^2 (Figure 2). Inside each sub-plot (1 m^2), the complete floristic composition was measured and the richness, percentage of bare soil, and foliage coverage was calculated using the Van der Maarel scale (1979). For the remaining plots, floristic composition was recorded and accumulated at the 1.000 m^2 plots. The unidentified plant species were collected and identified in the herbarium of the Universidad de Concepción (CONC), then they were classified according to their growth form in five categories: GM, graminoids (cyperaceous and juncaceous); GA, annual grasses (Poaceae); GP, perennial grasses (Poaceae); HA, annual herbs; HP; perennial herbs; AE, dwarf shrublands (prostrated or cushioned) and AM, medium shrublands (straight). Additionally, 4 edaphic samples were performed inside each plot, which were then mixed using a Core drill between 0 and 20 cm depth, obtaining a total of 16 cores of the soil's surface fraction. The samples were analyzed in the soil laboratory of INIA – Quilamapu. The physical variable evaluated was texture: clay (%), sand (%), lime (%) and retained humidity. The chemical variables evaluated were: pH in distilled water 1:25; extractable phosphorus determined using sodium bicarbonate 0.5M, pH 8.5, flow inj.; mineral nitrogen (NO_3 + NH_4) measured using potassium chloride 2N, flow inj.; organic matter; boron measured using sodium chloride 0.01M and colorimetry; sulfate SO_4 was determined using calcium phosphate 0.01M and turbidimetry; exchangeable cations Ca^{++}, Mg^{++}, Na^{+}, K^{+}, Al^{+++}, and total micro-nutrients of Zn, Fe, Mn and Cu.

Statistical Analysis

Data was analyzed with PC - ORD version 4.5 and SPSS Version 10.0 for Windows (Coakes and Steed, 2000). Differences in species richness due to grazing effects and edaphic conditions were evaluated utilizing MANOVA (Coakes and Steed, 2000). In this case, the number of response variables was fourteen (total species richness, native and exotic species richness, life forms, biomass and physicochemical parameters). Two factors were analyzed: vegetation type (shrublands vs. grass) and grazing treatment type (grazed and non-grazed). To determine the principal variations in the floristic composition and species abundance in shrublands and grasses in relation to grazing, a Detrended Correspondence Analysis (DCA; Ter Braak, 1986) was performed, using 480 sub-plots at the 1m^2 scale and 48 plots at the 1,000 m^2 scale. The similarity values of the species composition were compared between shrublands and grasses in grazed and non-grazed sites using the Jaccard's similarity coefficients ($S_j = a/a + b + c$) to compare the superposition of species lists, where S_j = Jaccard's similarity coefficient; a = number of species in the sample A and sample B; b = number of species in sample B but not in the sample A, and c = number of species in sample A but not in sample B (Krebs, 1999). Jaccard's coefficient was chosen over other similarity indexes because it treats all the species as of equal importance, considering that many plants have a coverage < 5% (Stohlgren *et al.*, 1999[a]).

RESULTS

Native and Exotic Species Richness and Coverage

For the 1200 hectares sampled, the total species richness was 89 species; 81 were native and 9 were exotic. This species richness represents 79% of the vascular flora of the Pali Aike National Park (Domínguez et al. 2004). The total species richness varied according to the spatial scale: for the 480 1-m^2 subplots, the total species richness found was 83 and for the 48 1000-m^2 plots, the richness was 89 species. The average richness of native species was significantly greater in *Festuca gracillima* grasses (16 species) then in shrublands of *Empetrum rubrum* (9 species) at the 1-m^2 scale (MANOVA, F= 691.7; $p<0.001$; Table 1). At the 1000 m^2 scale, the species richness was greater in grasses with 36 species than in shrublands with 18 species (MANOVA, F= 774.6; $p<0.001$; Table 2). Significant differences were found between grazed and non-grazed sites, where the average native species richness was higher in non-grazed sites with 14 species than in grazed sites with 10 species at the 1m^2 scale (MANOVA, F= 201; $p<0.001$; Table 1). At the 1000-m^2 scale, a similar pattern was found with the highest richness in the non-grazed sites, 31 native species and 22 species in grazed sites (MANOVA, F= 198; $p<0.001$; Table 2).

Differences were also found between grazing treatments. The highest mean richness for native species was found in the grasses in the non-grazed sites with 18.7 ± 0.3 species at the 1m^2 scale and 42.3 ± 0.8 at the 1000 m^2 scale (Table 1). The average foliage coverage was higher in non-grazed sites with a higher foliage coverage in grasses with 70% than in shrublands with 59% at the 1-m^2 scale (Table 3).

Table 1. Summarized MANOVA for plant community attributes: species richness, total (T), native (N) and exotic (E); and species richness by growth forms: AM, medium shrublands (straight); GA, annual grasses (Poaceae); GP, perennial grasses (Poaceae); GM, graminoids (cyperaceous and juncaceous); HA, annual herbs; HP, perennial herbs. Factors are: vegetation type (grassland vs. shrublands) and treatment (grazed vs. non-grazed) at the scale of 1 m^2 plots (n=480)

Factors	Richness	*F*	*P*	mean	mean	mean	mean
				Grasslands	Dwarf shrublands		
VEGETATION TYPE	T	1005.3	0.000	18.30	9.98		
	N	691.7	0.000	15.79	8.97		
	E	237.1	0.000	2.50	1.01		
	AE	65.0	0.000	2.17	1.59		
	AM	4.1	0.043	0.02	0.00		
	GA	90.4	0.000	0.27	0.00		
	GP	432.3	0.000	4.93	2.52		
	GM	119.2	0.000	0.81	0.30		
	HA	0.0	0.900	0.16	0.16		
	HP	477.9	0.000	9.95	5.40		
				Grazed	Non-grazed		
GRAZING	T	239.4	0.000	12.11	16.17		
	N	201.5	0.000	10.54	14.22		
	E	15.0	0.000	1.57	1.95		
	AE	74.7	0.000	2.19	1.57		
	AM	4.1	0.043	0.00	0.02		
	GA	10.7	0.001	0.18	0.09		
	GP	484.7	0.000	2.45	5.00		
	GM	0.1	0.790	0.56	0.55		
	HA	0.8	0.379	0.15	0.18		
	HP	111.3	0.000	6.58	8.77		
				Grazed Grasslands	Non-grazed Grasslands	Grazed Shrublands	Non-grazed Shrublands
VEGETATION TYPES * GRAZING	T	95.8	0.000	15.0	21.6	9.2	10.7
	N	64.5	0.000	12.9	18.7	8.2	9.8
	E	24.9	0.000	2.1	2.9	1.1	1.0
	AE	37.1	0.000	2.7	1.6	1.7	1.5
	AM	4.1	0.043	0.0	0.0	0.0	0.0
	GA	10.7	0.001	0.4	0.2	0.0	0.0
	GP	56.7	0.000	3.2	6.6	1.7	3.4
	GM	4.2	0.042	0.9	0.8	0.3	0.3
	HA	15.2	0.000	0.2	0.1	0.1	0.2
	HP	137.2	0.000	7.6	12.3	5.5	5.3

Table 2. Summarized MANOVA for plant community attributes: species richness, total (T), native (N) and exotic (E); and species richness by growth forms: AM, medium shrublands (straight); GA, annual grasses (Poaceae); GP, perennial grasses (Poaceae); GM, graminoids (cyperaceous and juncaceous); HA, annual herbs; HP, perennial herbs. Factors are: vegetation type (grassland vs. dwarf shrublands) and treatment (grazed vs. non-grazed) at the scale of 1000 m^2 plots (n=48)

Factors	Richness	*F*	*P*	*mean*	*mean*	*mean.*	*mean*
				Grassland	Dwarf shrublands		
VEGETATION TYPES	T	1179.6	0.000	41.79	19.42		
	N	774.6	0.000	35.54	17.83		
	E	375.0	0.000	6.25	1.58		
	AE	16.6	0.000	4.38	3.04		
	AM	3.1	0.083	0.17	0.00		
	GA	113.5	0.000	0.79	0.00		
	GM	32.8	0.000	1.42	0.71		
	GP	315.1	0.000	9.29	3.88		
	HA	24.0	0.000	0.88	0.25		
	HP	574.1	0.000	24.88	11.54		
				Grazed	Non-grazed		
GRAZING	T	182.1	0.000	26.21	35.00		
	N	198.2	0.000	22.21	31.17		
	E	0.5	0.493	4.00	3.83		
	AE	0.6	0.449	3.58	3.83		
	AM	3.1	0.083	0.00	0.17		
	GA	7.9	0.008	0.50	0.29		
	GM	0.1	0.738	1.08	1.04		
	GP	107.7	0.000	5.00	8.17		
	HA	0.1	0.745	0.58	0.54		
	HP	97.7	0.000	15.46	20.96		
				Grazed Grasslands	Non-grazed Grasslands	Grazed Shrublands	Non-grazed Shrublands
VEGETATION TYPES * GRAZING	T	40.1	0.000	35.33	48.25	17.08	21.75
	N	49.1	0.000	28.8	42.25	15.58	20.08
	E	1.9	0.174	6.50	6.00	1.50	1.67
	AE	1.0	0.314	4.42	4.33	2.75	3.33
	AM	3.1	0.083	0.00	0.33	0.00	0.00
	GA	7.9	0.008	1.00	0.58	0.00	0.00
	GM	25.5	0.000	1.75	1.08	0.42	1.00
	GP	9.0	0.004	7.25	11.33	2.75	5.00
	HA	2.7	0.109	1.00	0.75	0.17	0.33
	HP	63.0	0.000	19.92	29.83	11.00	12.08

Table 3. Abundance of native and exotic species (mean; ± SE) in dwarf shrublands and grasslands in grazed and ungrazed sites

	1 m^2 sub-plots †				1000 m^2 plots ‡	
	Native		Exotic		Native	Exotic
	Mean	Cover	Mean	Cover	Mean	Cover
Ungrazed sites						
Dwarf shrublands	9.8	59.3	1.0	0.30	20,08	1,7
	(0,33)	(0,7)	(0,08)	(0,02)	(0,60)	(0,26)
Grasslands	18.7	69.8	2.93	1.12	42,25	6
	(0,32)	(0,6)	(0,13)	(0,07)	(0,80)	(0,30)
Grazed sites						
Dwarf shrublands	8,2	52.0	1,1	0.49	15,58	1,5
	(0,19)	(0,5)	(0,06)	(0,04)	(0,34)	(0,23)
Grasslands	12.9	61.1	2.1	0.87	28,8	6.5
	(0,16)	(0,8)	(0,10)	(0,05)	(0,72)	(0,15)

† n = 480 sub-plots per site.
‡ n = 48 plots per site.

Nine exotic species were found, and they represented 50% of the introduced plants cited by the Pali Aike National Park (Domínguez *et al.,* 2004). The exotic species varied significantly between vegetation type, where the grasses had richer exotic species (2.5 species) than the shrublands (1.01 species) at the 1-m^2 scale (MANOVA, F= 237; $p<0.001$; Table 1).

At the 1000-m^2 scale, the greater exotic species richness was found in the grasses (6.25 species) rather than in shrublands (1.58 species) (MANOVA, F= 375; $p<0.001$). When comparing the average of exotic species between grazed and non-grazed sites, it was higher in grazed sites with 1.9 species in comparison with 1.6 species in non-grazed sites at the 1 m^2 scale (MANOVA, F= 15; $p<0.001$) (Table 1). At the 1000m^2 scale, no significant differences in exotic species richness was found between grazed and non-grazed sites (Table 2). Indeed, the greatest average exotic species richness between treatments was found in the grasses in non-grazed sites with 2.9 ± 0.1 species at the 1m^2 scale. At the 1000 m^2 scale, the greatest exotic species richness was found in grasses in grazed sites at 6.5 ± 0.1 (Figure 6; Table 3). Exotic species coverage was 1.12% in grasses and 0.30% in shrublands (Table 3).

Growth Forms

The growth forms varied significantly between shrublands and grasses at the 1 m^2 scales and were: (AE) dwarf shrublands, (AM) medium shrublands, (GA) annual grasses (Poaceae), (GP) perennial grasses (Poaceae), and (HP) perennial herbs (Table 4). The annual herbs (HA) were the only life form that did not present significant differences (MANOVA, F= 0.0;

p=0.900). For growth forms, there were no significant differences between grazed and non-grazed sites at the 1 m^2 scale for the annual herbs (MANOVA, F= 0.8; p=0.379) and the graminoids (MANOVA, F= 0.1; p=0.790) (Tables 1 and 4).

Table 4. Mean species richness and plant cover by growth form in grasslands and dwarf shrublands in grazed and non-grazed sites (n= 480 1m^2 plots). Growth forms: AE, dwarf shrublands (prostrated or cushioned); AM, medium shrublands (straight); GA, annual grasses (Poaceae); GP, perennial grasses (Poaceae); GM, graminoids (cyperaceous and juncaceous); HA, annual herbs; HP, perennial herbs; SD, bare soil cover.

Vegetation types	Grasslands				Dwarf shrubland			
Treatments	Grazed		Non-grazed		Grazed		Non-grazed	
Growth forms	Mean Sp.	Mean Cov.	Mean Sp.	Mean Cov.	Mean Sp.	Mean Cov.	Mean Sp.	Mean Cov.
AE	2,7	11,5	1,6	4,4	1,7	45,0	1,5	44,0
AM	0,0	0,0	0,0	0,3	0,0	0,0	0,0	0,0
GA	0,4	0,2	0,2	0,1	0,0	0,0	0,0	0,0
GM	0,9	1,6	0,8	3,7	1,7	0,1	0,3	0,1
GP	3,2	38,7	6,6	54,3	0,3	2,4	3,4	11,5
HA	0,2	0,1	0,1	0,0	0,1	0,1	0,2	0,1
HP	7,6	9,9	12,3	8,1	5,5	4,9	5,3	3,9
SD	38		29		47,5		40,4	

At the 1000 m^2 scale, the only growth form that was not significantly different between grasses and shrublands was the medium shrublands (MANOVA, F= 3.1; p=0.083; Table 2). There were no significant differences between grazed and non-grazed sites at this scale for the annual grasses, perennial grasses and perennial herbs (Table 2). The most consistent changes in growth forms at the 1 m^2 scale were in grasses in the grazed sites, where the dwarf shrublands increased their average coverage in 62% and the perennial herbs in 19%. In contrast, the graminoids decreased in 57%, the perennial grasses in 29%, and the medium shrublands disappeared (Table 4). The annual herbs and grasses had little participation in the grasses in the grazed sites, being represented only by two exotic species (*Aira proecox* and *Draba verna*) and one native (*Gerenium magellanicum*).

THE RESPONSE OF GRAZING-SENSITIVE SPECIES

An important group of grazing-sensitive plant species presented a consistent response to the grazing effect, being detected in non-grazed grasses of *Festuca gracillima* in Pali Aike National Park at the 1 m^2 scale, where 19 exclusive native species were found (*Junellia odonelli*, *Hordeum patagonicum*, *Calceolaria uniflora*, *Galium araucanum*, *Hamadryas delfinii*, *Adesmia villosa*, *Festuca pallescens*, *Bromus setifolius*, *Leucheria eriocephala*, *Huanaca acaulis*, *Senecio kingii*, *Draba funiculosa*, *Gamochaeta spiciformis*, *Vicia bijuga*, *Bromus catharticus*, *Leucheria purpurea*, *Carex macrosolen*, *Anemone multifida*) including the shrub species with conservation problems *Adesmia boronioides* according to Chile's red list (Benoit 1989). This species is absent in cattle raising areas because it is a foliage shrub favored by sheep (Domínguez personal observation). In contrast, only one exclusive native species (*Acaena magellanica*) was found in grazed sites.

ORDINATION ANALYSIS

With the DCA for 480 1 m^2 subplots (presence/absence data), the Eigenvalues for the first and second axis were 0.45 and 0.21, respectively. Axis I presented a continuous gradient with respect to the floristic species composition between grasses and shrublands independent of the grazing factor, which could indicate that grazing did not affect species richness at the local scale (Figure 3). At the same scale, but with abundance data, the Eigenvalues were 0.65 and 0.13 for the first and second axis, respectively. The first axis presented a clear bipolar segregation between both communities independent of the grazing effect (Figure 4). The DCA, for the 48 plots with absence/presence data at the 1000 m^2 scale, produced eigenvalues of 0.38 and 0.11 for the first and second axis, respectively. Axis I, of the DCA diagram, at the 1000 m^2 scale with absence/presence data, presented a clear segregation between grasses in grazed and non-grazed sites, which indicates that the grazing effect on floristic composition is more evident at the landscape scale (Figure 5).

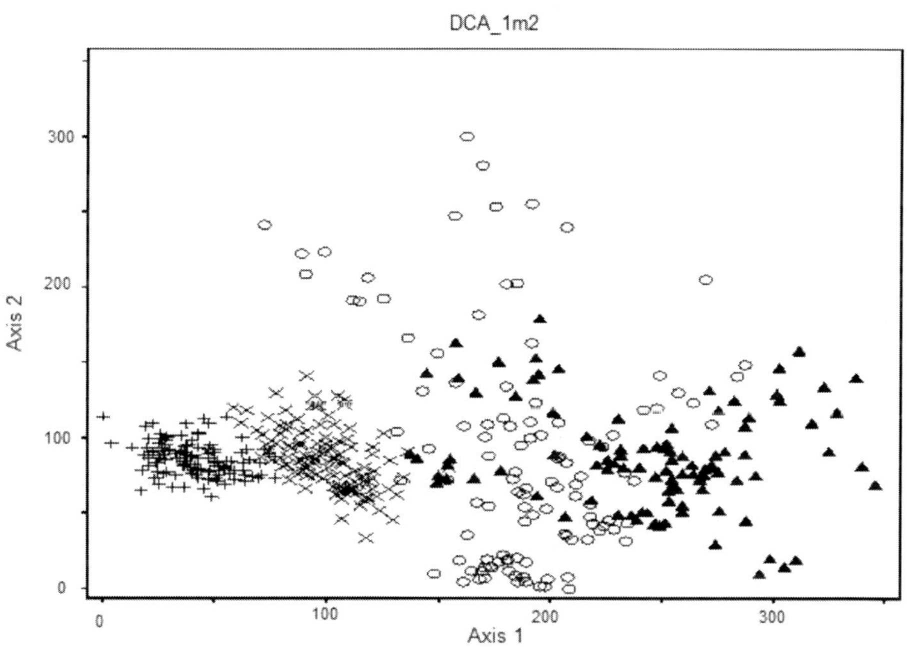

Figure 3. DCA ordination diagram for 480 subplots at the 1 m^2 scale using presence/abscence data for: (+) non-grazed grasslands of *Festuca gracillima* in excluded sites; (x) grasslands of *Festuca gracillima* in grazed sites; (○) dwarf shrublands of *Empetrum rubrum* in non-grazed sites; (▲) dwarf shrublands of *Empetrum rubrum* in grazed sites.

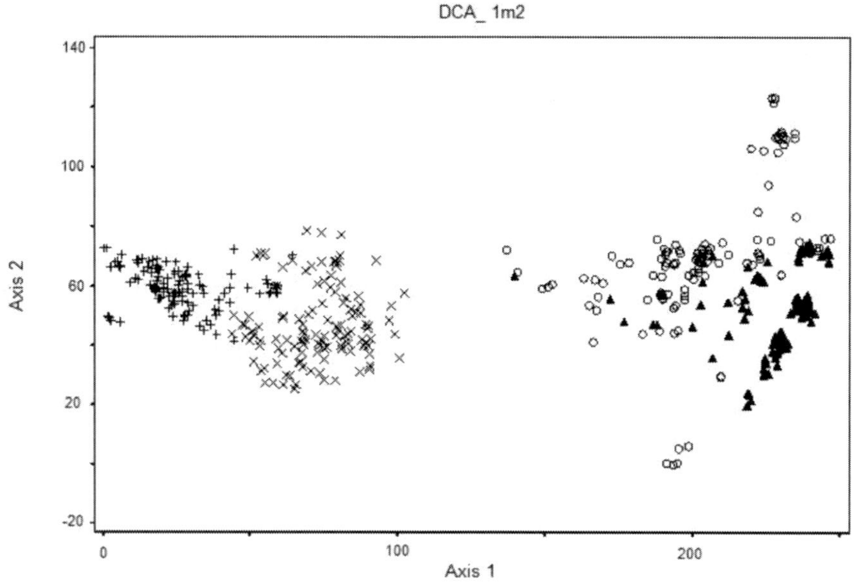

Figure 4. DCA ordination diagram for 480 subplots at the 1 m^2 scale using abundance data for: (+) non-grazed grasslands of *Festuca gracillima* in excluded sites; (x) grasslands of *Festuca gracillima* in grazed sites; (○) dwarf shrublands of *Empetrum rubrum* in non-grazed sites; (▲) dwarf shrublands of *Empetrum rubrum* in grazed sites.

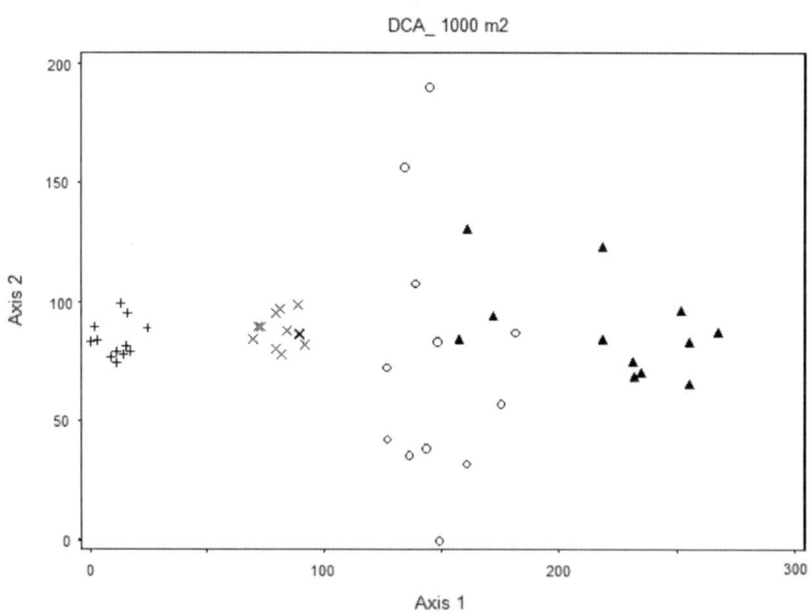

Figure 5. DCA ordination diagram for 48 subplots at the 1000 m^2 scale using presence/abscence data for: (+) non-grazed grasslands of *Festuca gracillima* in excluded sites; (x) grasslands of *Festuca gracillima* in grazed sites; (○) dwarf shrublands of *Empetrum rubrum* in non-grazed sites; (▲) dwarf shrublands of *Empetrum rubrum* in grazed sites.

Jaccard's Similarity Coefficient

The average of Jaccard's coefficient for the four contrast conditions indicated 24 % ± 0.005 of similarity in the floristic composition between grasses in grazed and non-grazed sites, followed by 21% ± 0.006 between shrublands in grazed and non-grazed sites at the 1 m^2 scale (Table 5). At the 1000 m^2 scale, the highest similarity coefficients were between grasses in grazed and non-grazed sites with 49 % ± 0.007, followed by 38% ± 0.009 between shrublands in grazed and non-grazed sites (Table 5).

Table 5. Jaccard Index (mean; ±SE) for the contrasts between the four studied conditions: S= dwarf shrublands, G= grassland, Gr = grazed, NGr = non-grazed at 1 m^2 sub-plots (n = 480) and 1000 m^2 plots (n = 48)

Scale	Conditions	GGr	GNGr	SGr	SNGr
1 m^2	GGr	1			
	GNGr	0.241 (0.005)	1		
	SGr	0.158 (0.005)	0.110 (0.003)	1	
	SNGr	0.179 (0.003)	0.135 (0.004)	0.219 (0.006)	1
	GGr	1			
1000 m^2	GNGr	0.492 (0.007)	1		
	SGr	0.276 (0.007)	0.319 (0.009)	1	
	SNGr	0.386 (0.009)	0.258 (0.008)	0.354 (0.013)	1

Soil Parameters

The physical and chemical soil characteristics presented significant differences between shrublands and grasses at the 1000 m^2 scale (Table 6), differentiating the shrublands soils for being thicker and with a larger proportion of sand (MANOVA, $F = 10.8$; $p=0.006$), more acidic (MANOVA, $F = 48$; $p = 0.000$), and with a high Al content (MANOVA, $F=120.7$; $p<0.001$; Table 6).

A superficial black-to-dark brown layer could be distinguished, incompletely humidified with a low N content in comparison with the grass soil (MANOVA, $F=55.52$; $p<0.001$) and with a higher gravimetric humidity percentage (MANOVA, $F= 7.4$; $p=0.0019$). The B layer had similar conditions for both vegetation types, which are characterized for being rocky of glacier origin.

The shrublands are principally affected by strong winds and by grazing, and both factors would be responsible for the high percentage of bare soil (Table 4). Significant differences were not found in organic matter content between shrublands and grasses (MANOVA, $F=0.0$; $p=0.846$). Only the texture and the soil's gravimetric humidity presented significant differences between grazed and non-grazed sites (Table 6).

Table 6. Summarized MANOVA for soil physicochemical parameters in grasslands and dwarf shrublands of grazed and non-grazed sites (n=16; in 1000 m^2 plots)

Factors	Variables	*F*	*P*	Mean	Mean	Mean	Mean
				Grasslands	Shrublands		
VEGETATION	P (ppm)	0.2	0.638	4.2	3.9		
TYPES	N (ppm)	55.5	0.000	12.2	3.2		
	pH (H_2O)	48.0	0.000	6.6	6.1		
	M.O. (%)	0.0	0.836	5.8	5.6		
	Ca (ppm)	18.1	0.001	8.7	5.3		
	Mg (ppm)	42.7	0.000	6.3	3.3		
	K (ppm)	9.3	0.010	0.8	0.5		
	Na (ppm)	6.4	0.027	0.4	0.7		
	Zn (ppm)	4.5	0.055	1.6	1.0		
	Fe (ppm)	23.7	0.000	117.5	219.1		
	Cu (ppm)	5.9	0.031	0.8	0.7		
	Mn (ppm)	0.0	0.990	3.2	3.2		
	Bo (ppm)	39.4	0.000	1.5	0.6		
	S (ppm)	0.5	0.484	1.3	1.5		
	Al (ppm)	120.7	0.000	23.4	90.1		
	Sand (%)	10.8	0.006	65.0	71.2		
	Silt (%)	16.1	0.002	24.4	17.4		
	Clay (%)	1.6	0.232	10.7	11.4		
	H. Grav. (%)	7.4	0.019	14.9	17.9		
				Grazed	Non-grazed		
GRAZING	P (ppm)	0.0	0.996	4.08	4.08		
	N (ppm)	0.3	0.583	8.08	7.40		
	pH (H_2O)	1.4	0.255	6.33	6.41		
	M.O. (%)	0.8	0.394	5.33	6.03		
	Ca (ppm)	0.8	0.403	6.67	7.35		
	Mg (ppm)	10.5	0.007	4.04	5.54		
	K (ppm)	3.0	0.108	0.60	0.74		
	Na (ppm)	1.4	0.259	0.61	0.49		
	Zn (ppm)	0.3	0.589	1.34	1.19		
	Fe (ppm)	4.0	0.070	147.56	189.04		
	Cu (ppm)	2.6	0.134	0.70	0.79		
	Mn (ppm)	0.0	0.861	3.29	3.14		
	Bo (ppm)	0.5	0.514	0.96	1.05		
	S (ppm)	1.2	0.299	1.50	1.30		
	Al (ppm)	0.3	0.603	58.38	55.13		
	Sand (%)	12.6	0.004	64.72	71.47		
	Silt (%)	21.0	0.001	24.87	16.87		
	Clay (%)	4.4	0.058	10.41	11.66		
	H. Grav. (%)	8.4	0.013	18.00	14.81		

				Grazed Grasslands	Non-grazed Grasslands	Grazed Shrublands	Non-grazed Shrublands
Factors	Variables	*F*	*P*	Mean	Mean	Mean	Mean
VEGETATION TYPES * GRAZING	P (ppm)	1.4	0.252	3.9	4.6	4.3	3.6
	N (ppm)	0.7	0.404	12.0	12.4	4.1	2.4
	pH (H_2O)	5.7	0.034	6.5	6.7	6.2	6.1
	M.O. (%)	5.1	0.044	6.3	5.2	4.4	6.8
	Ca (ppm)	4.5	0.055	9.2	8.2	4.2	6.5
	Mg (ppm)	0.1	0.728	5.5	7.1	2.6	3.9
	K (ppm)	0.4	0.539	0.8	0.8	0.4	0.6
	Na (ppm)	3.6	0.082	0.6	0.3	0.6	0.7
	Zn (ppm)	6.6	0.025	2.0	1.1	0.7	1.3
	Fe (ppm)	22.5	0.000	146.2	88.8	148.9	289.3
	Cu (ppm)	28.0	0.000	0.9	0.7	0.5	0.9
	Mn (ppm)	1.4	0.259	3.8	2.7	2.8	3.6
	Bo (ppm)	4.0	0.068	1.6	1.4	0.4	0.7
	S (ppm)	0.9	0.372	1.3	1.3	1.7	1.3
	Al (ppm)	0.2	0.630	26.5	20.3	90.3	90.0
	Sand (%)	2.1	0.173	60.2	69.7	69.2	73.2
	Silt (%)	0.7	0.407	29.1	19.6	20.6	14.1
	Clay (%)	4.4	0.058	10.7	10.7	10.2	12.7
	H. Grav. (%)	6.5	0.026	17.9	11.9	18.1	17.7

DISCUSSION

Our results indicate that grazing exclusion for 10 years has produced an increase in species diversity in grasses and dwarf shrublands at the local and landscape scale. However, it has not been sufficient to detect large environmental changes. These results suggest that the protection from grazing is an important factor for the reestablishment of grazing-sensitive native plant species, which is demonstrated by the presence of 23 species exclusively detected in the non-grazed sites within the National Park. These species can be indicating that the actual grazing rate is a key factor in the changes in diversity in the Austral Continental Patagonian steppe, more than nutrient availability in the soil as has been suggested for the steppe of Tierra del Fuego (Collantes *et al.,* 1989, 1999). The increase in species richness in non-grazed sites at the 1 and 1000 m^2 scales could be due to the absence of competitive exclusion or the suppression of dominant species provoked by past grazing in the Park as has been detected in other studies (Olff and Ritchie, 1998; Bullock and Pakeman, 1997).

Another factor that could influence species richness is the presence of open patches and clearing between the groups or cushions of grasses and herbs, most likely caused by past grazing prior to Park exclusion. This explication would explain the results observed in other studies that recognize that grazing could have indirectly affected the quality of microsites through modification of the cover of the grasses "tussocks" and the soil's physical and chemical characteristics (Dunkerley, 1997, Rambo and Faeth, 1999; Hunt, 2001; Cingolani *et al.,* 2005[a], 2005[b]). However, no chemical changes attributable to a grazing recess were detected in the present study. The changes detected were floristic in absence of grazing at the landscape scale, favoring the reestablishment of grazing-sensitive species, especially in grasses where 19 exclusive species were found. This result is in agreement with the observations of Chew (1982) for grasses in arid areas of southern Arizona (USA). It seems that in grazed sites, the sheep would be increasing the extinction rate of grasses and herbs in dwarf shrublands of *Empetrum rubrum*, which is confirmed by the low diversity found in the grazed sites in comparison with non-grazed sites.

The environmental differences detected in dwarf shrublands, especially the high acidity levels and Al concentration in comparison with grasses, could affect the colonization rate (Glen and Collins, 1992; Mendoza *et al.,* 1995). Similar levels for these environmental variables have been detected in Tierra del Fuego (Collantes, 1999). However, the species richness detected in these dwarf shrublands was greater than the value found in Tierra del Fuego by Collantes et al. (1989,1999). This greater richness is probably due to the lower acidity found on the continent. In contrast, a higher proportion of nutrients (N, P, K) was found for the grasses, and this characteristic would be determining the greater species diversity as has been suggested for other communities (Carrera et al. 2003, grasses in Northern Patagonia; Roem and Berendse 2000; Roem et al. 2002, heathland in Europe).

The exotic species had a participation of 11% for the total flora for the study area. A similar tendency has been detected in grazing prairies in Australia and Portugal (Gringulis *et al.,* 2001) , where the exotic species were only a minor component of the flora. The quantity of exotic species was significantly greater in grasses in comparison with shrublands at the 1 and 1000 m^2 scales. The analysis of exotic species in dwarf shrublands at the 1000 m^2 scale indicates a lower richness and coverage, probably due to the low soil fertility in dwarf shrublands of *E. rubrum* in comparison with the grasses of *F. gracillima.* This explanation is in agreement with the hypothesis that nutrient poor soils in semi-arid systems seem to be more resistant to invasions (Stohlgren *et al.,* 1999[a], 2001; Belnap and Phillips 2001; Gelbard and Harrison 2003). Considering that grasses in non-grazed sites are characterized for being nutrient rich and with a greater richness and foliage coverage of exotic species in comparison with nutrient poor sites, such as the dwarf shrublands of *E. rubrum.*

Of the 9 exotic species found in this study, all are related with the European immigrants who arrive in the 19th century the in Magellan's Strait area and introduced herbs and grasses from the Mediterranean regions, which coincide with the ecosystems of North America and Southern Australia, indicating a parallelism between the history of domestic grazing and the invasions in Austral Patagonia (Safford and Harrison, 2001). Only *Poa pratensis*, *Taraxacum officinale, Rumex acetosella, Cerastium arvense* and *Galium aparine* coincide with the exotic grasses and dwarf shrublands reported for cattle use in Tierra del Fuego by Collantes et al. (1999).

Our results indicate a negative response of perennial grasses at the actual grazing rate (0.76 UA/ha, SAG 2004), which is demonstrated by the diminishment in their richness and foliage coverage in grazed sites at the 1-m^2 scale. Studies performed in central Patagonia have demonstrated that grazing has a negative effect on the first establishment states of perennial grasses, which seems to be an important cause of the strong reduction of grass coverage and its eventual local extinction (Ares *et al.,* 2003; Bisigoto *et al.,* 2002; Bisigoto and Bertiller, 2004). Our results are in agreement with the observation made by Posse et al. (1996) in Tierra del Fuego, suggesting that the replacement of grasses is justified because the sheep find these more nutritive and palatable than the shrub species and perennial herbs (Posse *et al.,* 1996;

Anchorena, 2001). The most frequent shrub species (*Empetrum rubrum*, *Azorella monantha*, *Nardophyllum bryoides* and *Satureja darwinii*) are clearly not preferred by domestic cattle--demonstrated by the increase in their coverage in grazed sites. Among the woody ones, *Empetrum rubrum* is the only one used as an important resource for native herbivores (e.g., guanacos, ñandu and foxes) that eat their small leaves and especially their fruits.

It is probable that the native fauna is the principal dispersion agent for shrublands specie propagation. This characteristic could explain their success in rapid colonization of areas perturbed by domestic grazing, which are then scarcely occupied by the cattle as feeding sites as has been detected in Tierra del Fuego (Posse *et al.,* 1996), although native herbivores do use them as a competition-free food source, acquiring dominance in those sites (e.g., foxes and wild geese. Domínguez, unpublished). This could explain the surface area expansion of these dwarf shrublands, actually covering a surface area close to 24% of the 4 million hectares that form prairies of the austral Patagonian steppe ecosystem (Domínguez and Pauchard, unpublished).

It is important to note the scarce participation that the annual species had within the growth forms in the study area, finding only 3 species, represented by an exotic grass and herb and one herb native. These results do not agree with the tendency observed for the annual species in arid ecosystems (Jackson and Roy 1989). This difference is probably due to the late date of spring melting and the rapid drying that occurs between August and October, impeding annual species from developing their life cycle during the relatively short, environmentally unfavorable time period.

Our results show that the absence of domestic grazing in grasses of *Festuca gracillima* and shrublands of *Empetrum rubrum* during 10 years inside of a protected wild area permitted the increase of species richness by the reestablishment of 23 native plant species. This result indicates that current sheep grazing rate diminish grass diversity at the local scale (1 m^2), eliminating first the perennial herbs and grasses that are preferred by the herbivores. It appears, that grass communities of *Festuca gracillima* are highly resilient once the disturbance of grazing has ended, even permitting the reestablishment of woody species like the crawling bush *Junellia odonelli* absent in the steppe used for cattle grazing.

Protection from grazing inside the park has allowed the recovery of shrub species. For example, there are now species that are exclusive to the prairies inside the park such as *Adesmia boronioides,* which has been eradicated from the natural prairies that are grazed by domestic cattle. Protected populations in Chile are actually only found in the Torres del Paine National Park. This information permits the affirmation that the Pali Aike National Park conserves a remnant of the austral Patagonian steppe ecosystem that is characterized for its high diversity in comparison with those found in Tierra del Fuego.

The greater floristic heterogeneity in shrublands than in grasses could be driven by grazing. Woody shrub species increases and the perennial herbs and grasses decrease at higher grazing levels, as has been observed in other sectors of Patagonia (Márquez *et al.,* 2002; Bisigoto and Bertiller, 1997). The DCA at the landscape scale (1000 m^2) with presence/absence data shows a segregation model, probably due to the loss of infrequent species at the landscape scale, with the dominant species of each floristic type acquiring importance between grazed and non-grazed sites. Indeed, the excluded grasses are located on Axis I, to the left, followed by the grasses in the grazed sites, and the two types of shrublands, slightly mixed, are located to the right, both characterized by lower species richness.

Apparently, the alterations generated by grazing on the vegetal structure and composition are more evident at the landscape scale.

The greatest floristic similarity found between grasses in grazed and non-grazed sites was 49 % of similarity at the 1000 m^2 scale and 24% at the 1 m^2 scale. This would confirm that: 1) grazing reduces plant species richness, but does not generate irreversible changes in structural terms that can result in the conversion from *Festuca gracillima* grasses to *E. rubrum* shrublands as has been suggested for the prairies that are located in poor environments and that have quite a lot of dwarf shrublands in its composition as occurs in Tierra del Fuego (Posse *et al.,* 2000; Borrelli and Oliva, 2001; Cingolani *et al.,* 2005), 2) the floristic similarity will depend on the measurement scale, where the greater the scale, the floristic similarity is greater, and 3) grasses and shrublands are floristically and environmentally different.

We identified significant differences between shrublands and grasses in terms of composition and structure between grazed and non-grazed sites. However, no significant differences in the edaphic variables (e.g., pH y Al) were found between grazed and non-grazed sites. It seems that 10 years was not sufficient time to detect edaphic changes generated by the absence of grazing. However, the soil characteristics associated to the shrublands of *Empetrum* were similar to those observed in Tierra del Fuego: acid soils, the N levels, and high Al concentrations (Collantes *et al.,* 1989). Surely, the low native and exotic species richness detected in those shrublands in grazed and non-grazed sites could be due to the presence of Al, which has been cited as a fertility gradient indicator due to its toxicity in shrublands of *E. rubrum* (Collantes et al. 1989).

But, our data indicate that a slight increase in species richness in dwarf shrublands inside the Pali Aike National Park. This could be attributed to the significant differences detected in the texture and the soil's gravimetric humidity between grazed and non-grazed sites. These differences can be explained by the movements and compacting provoked by cattle treading outside the park. These small disturbances could generate different micro-environmental conditions at the local scale, possibly affecting the seed bank as has been reported by Márquez et al (2002). This could explain the greater species richness in non-grazed sites. However, no significant difference in richness and seed bank density was found between the grazed and non-grazed sites as has been reported for the Province of Cordoba, Argentina (Márquez et al. 2002). Additionally, it is unlikely that the grazing could trigger changes in soil characteristics as has been suggested in studies in Tierra del Fuego (Collantes et al. 1989; Mendoza et al 1995).

Changes detected in non-grazed sites represent a rapid response to the relief of an inadequate grazing rate, which is utilized in the majority of the cattle ranches in austral Patagonian region. This grazing pressure appears to be unfavorable for species diversity conservation and the prevalence of growth forms that characterize an ecosystem such as *Festuca gracillima* tussocks.

Therefore, the extensive management of Patagonian grasses and shrublands should aim to reduce grazing pressure in order to avoid prairie degradation and to generate sustainable changes in diversity that is compatible with cattle and environmental activity. Our results show that it is possible, in a relatively short period of time, restore at least partially the diversity of these unique grasslands and shrublands.

ACKNOWLEDGMENTS

The authors thank Ana Cingolani and Lohengrin Cavieres for their valuable comments on earlier versions of this manuscript. We also thank Prof. Urcecino Gonzáles for his valuable comments and statistical support, Nilo Covacevich for the logistic support in the field, and Centro de Estudios del Cuaternario de Fuego-Patagonia y Antártica (CEQUA) for the financial support of E. Dominguez's Master's study. Aníbal Pauchard is funded by ICM P05-002 and CONICYT PFB-23.

REFERENCES

Adler, P.B., Milchunas, D.G., Sala, O.S., Burke, I.C., and Lauenroth, W.K. (2005). Plant traits and ecosystem grazing effects: comparison of U.S sagebrush steppe and Patagonian steppe. *Ecol. Appl. 15*, 774-972.

Anchorena, J., and Cingolani, A. (2002). Identifyting habitat types in a disturbed area of the forest-steppe of vegetation. *Oecologia 128*, 465-479.

Ares, J.O., Bertiller, M.B., and Bisigoto, A.J. (2003). Modeling and measurement of structural changes at a landscape scale in dryland areas. *Ecol. Model. Assessment 8*, 1-13.

Baetti, C., Borrelli, P., and Collantes, M. (1993). Sitios glaciares y fluvioglaciares del Norte de Tierra del Fuego. En: *Secuencias del deterioro en distintos ambientes patagónicos. Su caracterización mediante el modelo de estados y Transiciones* Eds. Ludepa SME. Bariloche: 1003-109.

Bakker, J.P. (1999). Constraints in the restoration of ecological diversity in grassland and heathland communities. *J. Veg. Sci. 14*, 63-69.

Bakker, J.P. (1989). The impact of grazing to plant communities. In: Wallis De Vries M.F., Bakker J.P. and Van Wieren S.E. (eds.) *Grazing and conservation management.* Kluwer, Dordrecht, the Netherland, 137-184.

Belnap, J., and Phillips, S.L. (2001). Soil biota in an ungrazed grassland: response to annul grass (*Bromus tectorum*) invasion. *Ecol. Appl. 11*, 1261-1275.

Belsky, A.J. (1992). Effects of grazing, competition, disturbance and fire on species composition and diversity in grassland communities. *J. Veg. Sci. 3*, 187-200.

Benoit, I.L. (ed). (1988). *Libro Rojo de la Flora Terrestre de Chile*. CONAF, Santiago, Chile, 157.

Bokdam, J., and Gleichman, J.M. (2000). Effects of grazing by free-ranging cattle on vegetation dynamic in a continental north-west European heathland. *J. Appl. Ecol. 37*, 415-431.

Bisigoto, A.J., and Bertiller, J.M. (2004). Temporal and micro-spatial patterning of seedling establishment, consequences for patch dynamics in the southern Monte, Argentina. *Plant Ecol. 174*, 235-246.

Bisigoto, A.J., Ares, J.O., and Bertiller, M.B. (2002). Assessment of pristine vegetation structure in semiarid shrublands base on spatial explicit modeling. *Phytocoenologia 32*, 581-342.

Bisigoto, A.J., and Bertiller, M.B. (1997). Grazing effects on patchy dryland vegetation in northern Patagonia. *J. Arid Environ. 36*, 639-653.

Bertiller, M.B. (1998). Spatial patterns of the germinable soil seed bank in northern Patagonia. *Seed Sci. Res. 8*, 39-45.

Borreli, P., and Oliva, G. (2001). Efectos de los animales sobre los pastizales. Cap. 4.Pp. 101-130. En: *Ganadería Ovina Sustentable en la Patagonia Austral.* Borrelli, P. y G. Oliva Editores. INTA. Reg. Patagonia Sur. 272.

Breeskow, A.M., Elissalde, N.O., and Rostagno, C.M. (1995). Ecosystem changes associated with grazing intensity on the Punta Ninfas rangelands of Patagonia, Argentina. *J. Range Manage. 48*, 517-522.

Bullock, J.M., and Pakeman, R.J. (1997). Grazing of lowland heath in England: management methods and their affects on heathland vegetation. *Biol. Conserv. 79*, 1-13.

Carrera, A.L., Bertiller, M.B., Sain, C.L., and Mazzarino, M.J. (2003). Relationships between plant nitrogen conservation strategies and the dynamics of soil in the arid Patagonian Monte, Argentina. *Plant and Soil 255*, 595-604.

Chew, R.M. (1982). Changes in herbaceous and suffrutescent perennials in grazed and ungrazed desertified grassland in southastern Arizona. 1958-78. *Am. Midl. Nat. 180*, 159-169.

Cingolani, AM., Posse, G., and Collantes, M. (2005). Plant functional traits, herbivore selectivity and response to sheep grazing in Pagonian steppe grasslands. *J. Appl. Ecol. 42*, 50-59.

Cingolani, A.M., Noy-Meir, I., and Díaz, S. (2005). Grazing effects on rangeland diversity: a synthesis of contemporary models. *Ecol. Appl 15*, 757-773.

Cingolani, A.M, Cabido, MR., Renison, D., and Solís Neffal, V. (2003). Combined effects of environment and grazing on vegetation structure in Argentine granite grasslands. *J. Veg. Sci. 14*, 223-232.

Cingolani, A.M., Anchorena, J., and Collantes, M. (1998). Landscape heterogeneity and long-term animal production in Tierra del Fuego. *J. Range Manage. 51*, 79-87.

Coakes, S.J., and Steed, L.G. (2000). *SPSS analysis without anguish.* New York: Wiley. 283 pp.

Collantes, M.B., Anchorena, J., and Koremblit, G. (1989). A soil nutrient gradient in Magellanic *Empetrum* heathlands. *Vegetatio 80*, 183-193.

Collantes, M.B., Anchorena, J., and Cingolani, A.M. (1999). The steppes of Tierra del Fuego: floristic and growthform patterns controlled by soil fertility and moisture. *Plant Ecol.148*, 61-75.

Davis, M.A., Grime, J.P., and Thompson, K. (2000). Fluctuating resources in plant communities: a general theory of invasibility. *J. Ecol. 88*, 528-534.

Díaz, S., Acosta, A., and Cabido, M. (1994). Community structure in montane grassland of central Argentina in relation to land use. *J. Veg. Sci. 5*, 483-488.

Domínguez, E., Marticorena, C., Elvebakk, A., and Pauchard, A. (2004). Catálogo de la flora vascular del Parque Nacional Pali Aike. XII Región, Chile. *Gayana Bot. 61*, 67-72.

Dukerley, D.L. (1997). Banded vegetation: survival under drought and grazing pressure based on a simple cellular automaton model. *J. Arid Environ. 35*, 419-428.

Endlicher, W., and Santana, A. (1988). El clima del sur de la Patagonia y sus aspectos ecológicos. Un siglo de mediciones climatológicas en Punta Arenas. *An. Inst. Patagonia*, Serie *Cs. Nat. 26*, 119-128.

Gelbard, J.L., and Harrison, S. (2003). Roadless habitats as refuges for native grassland: interactions with soil, aspect, and grazing. *Ecol. Appl. 13*, 404-415.

Glen. S.M., and Collins S.L. (1992). Effects of scale and disturbance on rates of immigration and extinction of species in prairies. *Oikos 63*, 273-280.

Grime, J.P.(1973). Competitive exclusion in herbaceous vegetation. *Nature 242*, 34-47.

Gringulis, K., Sheppard, A.W., Ash, J.E., and Groves, R.H. (2001). The comparative demography of the pasture weeb *Echium plantagineum* between its native and invaded ranges. *J. Appl. Ecol. 38*, 281-290.

Güsewell, S., Kathleen, M.B., Roem, J.W., and Bedford, B.L. (2005). Nutrient limitation and botanical diversity in wetland: can fertilization raise species richness? *Oikos 109*, 71-80.

Holmgren, M., Avilés, R., Sierralta, L., Segura, A., and Fuentes, E. R. (2000). Why have European herbs so successfully invade the Chilean matorral? Effects of herbivory, soil nutrients, and fire. *J. Arid Environ. 44*, 197-211.

Huisman, J., Grover, J.P., Van Der Wal, R., and Van Andel, J. (1999). Competition for light, plant species replacement, and herbivore abundance along productivity gradients. Herbivores: *Between plants and predators* (eds.) H. Olff, V.K Brown and R.H. Drent. 239-269 pp.

Hulme, P.D., Pakeman, R.J., Torvell, L., Fisher, J.M., and Gordon, I.J. (1999). The affects of controller sheep grazing on the dynamic of upland Agrostis-Festuca grassland. *J. Appl. Ecol. 36*, 886-900.

Hunt, L.P. (2001). Heterogeneous grazing causes local extinction of edible perennial shrubs: a matrix analysis. *J. Appl. Ecol. 38*, 238-252.

Jackson, L.E., and Roy, J. (1989). Comparative ecology of annual grasses: native versus Californian Habitats. Pp. 81-91. (eds.) L.F. Huenneke and H.A. Mooney. *Grassland structure and funtion: California annual grassland.* Kluwer Academic Press, Dordrecht, Germany.

Koremblit, G., and Force, J.A. Ley. 1991. Contribución al estudio agroclimático del norte de Tierra del Fuego (Argentina). *An. Inst. Patagonia,* Serie *Sc. Nat.* (Chile) *20*, 125-1354.

Krebs C.J. (1999). *Ecological methodology* . 2nd edn. Addison-Welsey Education Publishers, Inc, Menlo Park, California, USA, (under Benjamin/Cummings Imprint).

Lauenroth, W.K., and Agujera, M.O., (1998). Plant-plant interactions in grasses and grasslands. Pp. 209-230. In: Cheplick G.P. (ed.), *Population biology of grasses.* Cambridge University Press. Cambridge, UK.

Mackey, R.L., and Curie, D. J. (2001). The diversity disturbance relationship: is it generally strong and peaked? *Ecology 82*, 3479-3492.

Márquez, S., Funes, G., Cabido, M., and Pucheta, E. (2002). Efectos del pastoreo sobre el banco de semillas germinable y la vegetación establecida en pastizales de montaña del centro de Argentina. *Rev. Chil. Hist. Nat. 75*, 327-337.

Mendoza, R.E., Collantes, M.B., Anchorena, J., and Cainzos, S. (1995). Effects of liming and fertilization on forage yield and vegetation in dry heath from Tierra del Fuego. *J. Plant Nutrition 18*, 402-420.

Milchunas, D.C., Sala, O.E., and Lauenroth, W. (1988). A generalized model of the effects of grazing by large herbivores on grassland community structure. *Am. Nat. 132*, 87-106.

Olff, H., and Ritchie, E. (1998). Effects of herbivores on grassland plant diversity. *Trends Ecol. Evol. 13, 261-265.*

Oliva, G., Cibils, A., Borrelli, P., and Humano, G. (1998). Stable states in relation to grazing in Patagonia. A 10 year experimental trial. *J. Arid Environ. 40*, 113-131.

Osen, Y., Perevolotsky, A., and Kigel, J. (2002). Grazing effects on diversity of annual plant communities in a semi-arid rangeland: interactions with small-scale spatial and temporal variation in primary productivity. *J.Ecol.* 90, 936-946.

Pisano, E. (1977). Fitogeografía de Fuego-Patagonia Chilena. I. Comunidades vegetales entre las latitudes 52º S y 56º S. *An. Inst. Patagonia*, Punta Arenas Chile *8*, 121-250.

Posse, G., Anchorena, J., and Collantes, M.B. (2000). Spatial micro-patterns in the steppe of Tierra del Fuego induced by sheep grazing. *J. Veg. Sci. 11*, 43-50.

Posse, G., Anchorena, J., and Collantes, M.B. (1996). Seasonal diets of sheep in the steppe region of Tierra del Fuego, Argentina. *J. Range Manage. 49*, 24-30.

Prouxlx, M., and Mazumber, A. (1998). Reversal of grazing impact on plant species richness in nutrient-poor vs. nutrient-rich ecosystem. *Ecology 79*, 2581-2592.

Pykälä, J. (2004). Cattle grazing increases plant species richness of most species trait groups in mesic semi-natural grasslands. *Plan Ecol. 175*, 217-226.

Rambo, J.L., and Faeth, S.H. (1999). Effect of vertebrate grazing on plant and insect community structure. *Conserv. Biol. 13*, 1047-1054.

Roem, W.J., Klees, H., and Berendse, F. (2002). Effects of nutrient addition and acidification on plant species diversity and seed germination in heathland. *J. Appl. Ecol. 39*, 937-948.

Roem, W.J., and Berendse, F. (2000). Soil acidity and nutrient supply ratio as possible factors determining changes in plant species diversity in grassland and heathland communities. *Biology Conservation 92*, 151-161.

Roques, K.G., O'Connor, T.G., and Watkinson, A.R. (2001). Dynamics of shrub encroachment in an African savanna: relative influences of fire and herbivory, rainfall and density dependence. *J. Appl. Ecol. 38,* 268-280.

Saez, C. (1995). *Fertilidad de los suelos de la Región de Magallanes.* Universidad de Magallanes 5-18.

Safford, H.D., and Harrison, S.P. (2001). Grazing and substrate interact to affect native vs. exotic diversity in roadside grasslands. *Ecol. Appl. 11*, 1112-1122.

Servicio Agrícola y Ganadero (SAG). (2004). *El pastizal de Magallanes, guía uso, condición actual y propuesta de seguimiento para determinación de tendencia.* (eds.) Servicio Agrícola Ganadero, XII Región de Magallanes. 127 pp.

Scholes, R.J and Archer, S.R. (1997). Tree-grass interactions in savannas annual review of *Ecology and Systematic 28*, 517-544.

Stoghlgren, T.J., Otsuki, Y., Villa, C.A., Lee, M., and Belnap, J. (2001). Petterns of plant invasion: a case example in native species hotspots and rare habitats. *Biol. Invasions 3*, 37-50.

[a]Stohlgren T.J., Binkley, D., and Chong, G.W. (1999). Exotic plant species invade hot spots of native plant diversity. *Ecol. Monogr. 69*, 25-46.

[b]Stohlgren, T.J., Schell, L.D., and Heuvel, B. V. (1999). How grazing and soil quality affect native and exotic plant diversity in rocky mountain grasslands. *Ecol. Appl. 9*, 45-64.

Stohlgren, T.J., Bull, K.A., and Otsuki, Y. (1998). Comparison of rangeland vegetation sampling techniques in the Central Grasslands. *J. Range Manage. 51*, 164-172.

Ter Braak, C.J.F. (1986). Canonical correspondence analysis: a new eigenvector technique for multivariate direct gradient analysis. *Ecology 67*, 1167-1179.

Tilman, D. (1988). *Plant strategies and the dynamic and structure of plant communities.* Princeton University Press, Princeton, New Jercy.

Van der Maarel, E. (1979). Transformation of cover-abundance values in phytosociology and its affects on community similarity. *Vegetatio 39*, 97-114.

Walter, H., and Box, E.O. (1983). Climate of Patagonia. Pp. 432-435. In: West, N.E. (eds.), *Ecosystems of the world*, Vol. 5. Temperate deserts and semi-deserts. Elsevier Scient. Publi. Co, Amsterdam. Oxford.

Watkinson, A.R., and Ormedod, S.J. (2001). Grassland, grazing and biodiversity: editors' introduction. *J. Appl. Ecol. 38*, 233-237.

In: Grazing Ecology: Vegetation and Soil Impact
Editor: José Ramón Arévalo
ISBN 978-1-62100-436-3

Chapter 6

THE ROLE OF THE NUTRITIONAL AND ANTI-NUTRITIONAL CHARACTERISTICS IN GRAZING MANAGEMENT PROGRAMS TO CONTROL *CENTAUREA SOLSTITIALIS*

Pelayo Álvarez*
Defenders of Wildlife, Sacramento ,CA. USA

ABSTRACT

Yellow starthistle (*Centaurea solstitialis*) is the most widely distributed invasive weed in California. It has infested an estimated 14.3 million acres and is present in 56 of the 58 counties. Yellow starthistle infestations cost the State of California millions of dollars in losses and impacts both natural and agricultural systems. Grazing is one of the available methods for the control of yellow starthistle and is already in use by private landowners and several land management agencies in California. However, starthistle produces sesquiterpene lactones that may cause adverse effects to large herbivores ranging from lethal to digestive discomfort. In addition the sesquiterpene lactones present in yellow starthistle can accumulate in animal tissues or secreted in milk of livestock used to control this weed. The presence of sesquiterpene lactones in yellow starthistle could hinder the implementation of grazing management programs to control this weed.

This chapter comprises a combination of field trials and detailed laboratory work aimed at determining the seasonal variability of nutritional and anti-nutritional characteristics of yellow starthistle and the effects of ingestion of yellow starthistle in sheep. The following conclusions were drawn from this chapter:

1. Yellow starthistle has good nutritional value for sheep. Feed intake and digestibility of yellow starthistle meet the nutritional requirements for maintenance for sheep.
2. The sesquiterpene lactones present in yellow starhistle do not have anti-quality properties and do not affect the health of sheep grazing this plant.

* Email: palvarez@defenders.org. Phone: 916-313-5800 x 107

Based on the information gathered in this chapter, sheep grazing can be an effective tool in the control of yellow starthistle without deleterious effects to the health and well-being of the sheep.

Keywords: Yellow starthistle, invasive species, management of exotics, grazing, toxicity

INTRODUCTION

Yellow starthistle (*Centaurea solstitialis*) is the most widely distributed invasive weed in California. It has infested an estimated 14.3 million acres and is present in 56 of the 58 counties (Pitcairn *et al.,* 2006). Yellow starthistle infestations cost the State of California millions of dollars in losses and impacts both natural and agricultural systems (DiTomaso, 2000). In natural systems, yellow starthistle invasions reduce wildlife habitat and forage quality, and decrease biodiversity by displacing native plant species (Sheley and Larson, 1994). In addition, yellow starthistle infestations fragment vulnerable plant and animal habitat (Callihan *et al.,* 1995). Yellow starthistle also impacts agriculture negatively by interfering with livestock grazing and reducing overall productivity of rangelands (Barry, 1995).

Several characteristics make yellow starthistle a successful invader in California rangelands. Yellow starthistle has an extended germination period closely related to rainfall events (Benefield *et al.,* 2001). The plant grows initially as a rosette and produces a robust tap root that can reach a depth of 6 feet by early summer which allows the plant access to moisture long after the end of the rainy season (Roche *et al.*, 1994; DiTomaso *et al.*, 2003). In the Sacramento Valley of California, bolting occurs in the spring to early summer with flowering occurring from late May to September (Benefield *et al.*, 2001). Mature plants can produce 50-100 million seeds per acre (20-40 million seeds/ha) and are able to live in the soil from 3 to 10 years (DiTomaso *et al.*, 1999; Maddox, 1981). By delaying the completion of its life cycle relative to the annual grasses, yellow starthistle is able to reduce competition at the reproductive stage (Roche *et al.*, 1994; Thomsen *et al.*, 1993). However, this ability to outcompete the herbaceous annuals makes yellow starthistle especially vulnerable to grazing by increasing its palatability relative to the already dried annuals (Thomsen *et al.*, 1993).

Grazing is one of the available methods for the control of yellow starthistle (DiTomaso, 2000) and is already in use by private landowners and several land management agencies in California. Domestic sheep (*Ovis aries*) consume yellow starthistle in the bolting stages causing a significant reduction in seed production (Thomsen *et al.* 1989). However, starthistle produces plant secondary compounds including alkaloids (Mamedov, 1956), terpenoids (Buttery *et al.*, 1986), and sesquiterpene lactones (Merrill and Stevens, 1985). Due to their anti-quality properties, secondary compounds may cause adverse effects to large herbivores that range from lethal to digestive discomfort (Launchbaugh *et al.*, 2001). When ingested by horses, yellow starthistle causes a neurological disorder of the brain called nigropallidal encephalomalacia that causes immobility of the facial musculature and impaired eating and drinking followed by hypokinesia and death (Cordy, 1978). The specific etiology of the disease has not been determined but sesquiterpene lactones (Stevens *et al.,* 1990; Akbar *et al.,* 1995) have been implicated. Hymenovin, a sesquiterpene lactone found in bitterweed (*Hymenoxys odorata*) (Ivie *et al.*, 1975), *Hymenoxys richardsonii* var. floribunda (Ivie *et al.*, 1976) and *Dougaldia hoopesii* (Ivie *et al.*, 1976; Hill *et al.*, 1977) is toxic to sheep. The same sesquiterpene lactone is present in Orange sneezeweed (*Helenium hoopesii*) and causes a disease named "spewing sickness" (Hesker, 1982) in sheep.

Other sesquiterpene lactones are not toxic but have anti-nutritional properties. When bitterweed (*Hymenoxis odorata*) is added to the diet of sheep voluntary intake decreases (Calhoun et al. 1981). Cnicin a sesquiterpene lactone present in spotted knapweed (*Centaurea maculosa*) inhibited ruminal microbial gas production in vitro (Olson and Kelsey, 1997).The sesquiterpene lactones present in yellow starthistle could hinder the implementation of grazing management programs to control this weed.

The seasonal variation in nutritional composition and the potential anti-nutritional properties of yellow starthistle in sheep are unknown. Anti-quality factors may adversely affect production parameters and the welfare of those animals used in grazing management programs to control invasive species. In addition, nutritional composition and anti-nutritional properties of plants are essential in determining diet selection by large herbivores (Freeland and Janzen, 1974; Provenza and Balph, 1987; Foley *et al.*, 1999; Dearing *et al.*, 2005). Knowledge of these properties is essential for evaluating the efficacy of sheep grazing as a tool to control yellow starthistle. The purpose of this chapter was to determine the nutritional composition, intake and *in-vitro* and *in-vivo* digestibility of *Centaurea solstitialis* and to evaluate the effect of the ingestion of secondary compounds present in bolting yellow starthistle in sheep.

Materials and Methods

Study Site

Field studies were conducted in 2004 at two sites at the campus of the University of California, Davis (38.53 ° N, 121.78° W) and 18 m elevation, in Yolo County, California, USA. The soil at the study site is Yolo Silty Loam and is a deep alluvial soil characteristic of the northern Central Valley. The climate at the study site is typically Mediterranean with hot, dry summers and cool, wet winters. Average annual rainfall is 491 mm and most precipitation occurs between October and May.

Plant Material

Site A: Seeds of *Centaurea solstitialis* collected in the Fall of 2003 were broadcast-seeded and watered twice to help establishment. Commercial urea was applied once at a rate of 6 lbs per acre. Plants were collected from each site during each of 3 phenological stages: rosette, bolting, and flowering. Approximately 350 g in dry matter were collected by manually clipping several plants at the same growth stage at ground level.

Site B: Plants were collected from each site during each of 3 phenological stages: rosette, bolting, and flowering. Approximately 350 grs in dry matter were collected by manually clipping several plants at the same growth stage at ground level.

Animal Management

The experiments were conducted with Targhee ewes (average initial weight 46 ± 3 kg). All six animals were surgically fitted with rumen cannulae. To determine intake and *in vivo* digestibility, animals were housed in individual metabolism crates and fed a diet of 100% yellow starthistle for four days, following and adaptation period of 17 days. Fresh plant material was harvested every morning from site A and fed to the sheep twice a day in the morning and in the evening. After the feeding trial, animals where moved to a pen and fed yellow starthistle supplemented with alfalfa hay until starthistle reached the flowering stage (July 15th). Animals were fed yellow starthistle for a total of 65 days (May 15th–July 19th). The experimental protocol was reviewed and accepted by Institutional Animal Care and Use Committee at the University of California, Davis.

Laboratory Analyses

Feed intake and *in vivo* diet digestibility were measured by total fecal collection. Daily sub samples from the fresh feed offered and from composited total feed refusals for each individual animal were weighed and taken for DM determination and further analyses. Samples of feed on offer, refusals and feces were dried in a forced-air oven (55°C for 72 h) and ground to pass a 1 mm-screen in a Wiley mill (Arthur A. Thomas, Philadelphia, PA) and stored until analysis. Samples were analyzed for organic matter (OM), ash, and crude protein (CP) (N x 6.25) according to the methods described in AOAC (1990). Neutral detergent fiber (NDF) and acid detergent fiber (ADF) were determined following the method of Van Soest (1991).

In Vitro Gas Production

In vitro gas production was obtained using the technique described by Menke and Steingass (1988). Approximately 200 mg of sample were weighed in duplicate into a 100 ml graduated glass syringe. Pistons were lubricated with vaseline and inserted into the syringes. Buffer and mineral solution (Makkar *et al.,* 1995) were prepared and placed in a water bath at 39°C under continuous flushing with CO_2. Rumen fluid was collected before the morning meal from two rumen-fistulated, non-lactating, non-pregnant Holstein cows (about 800 kg body weight) fed grass hay diet. Rumen fluid was pumped with a manually operated air pump from the rumen into pre-warmed thermos flasks. The rumen fluid from the two cows was mixed and filtered through four layers of cheesecloth and then flushed with CO_2. The CO_2 flushed rumen fluid was added to the buffered mineral solution (1:3 v/v), which was maintained in a water bath at 39°C, and mixed. Buffered rumen fluid (30 ml) was dispensed into each syringe containing ground browse samples. The syringes were immediately placed in a water bath at 39°C. Three syringes containing 30 ml inoculum served as blanks (no sample). Incubations were performed by duplicated. The gas production at 24 h was corrected for day-to-day variation in the activity of rumen liquor using the Hohenheim hay standard. The syringes were gently shaken every 2 h, and the incubation terminated after recording the

72 h gas volume. Gas production (Gp) was recorded before incubation (0) and 2, 4, 6, 8, 10, 12, 24, 26, 28, 30, 32, 36, 48, 52, 54, 56, 60 and 72 h. Total gas values were corrected for blank incubations, and reported gas values are expressed per g of DM. Volume of gas produced after 24 h of incubation was used as an index of digestibility and energy feed value.

Digestibility and energy content were calculated by the following relationships:

$$\text{OMD (\%)} = 14.88 + 0.889\, Gp + 0.45\, \text{CP} + 0.0651\, XA.$$

$$\text{ME (Mcal/kg DM)} = 2.2 + 0.136 * Gp24 + 0.057 * CP.$$

where OMD is the organic matter digestibility; ME = metabolizable energy; *CP*= crude protein in %, *XA* = ash in %; and *Gp*= the net gas production in ml from 200 mg dry sample after 24 h of incubation (Menke and Steingass, 1988).

To measure the toxicity of yellow starthistle to sheep, blood samples were collectedvia indwelling jugular catheters (Micro-Renathane, Braintree Scientific, Braintree, MA).into vacuum glass tubes without anticoagulant before feeding yellow starthistle (control) and in the last day of the feeding trials. After clotting, the serum was separated by centrifugation (4,500 rpm for 15 minutes) transferred into plastic vials and stored in a freezer at -70 ° C for later analyses. The UC Davis Veterinary Diagnostic Laboratory used a Roche Cobas Mira S Plus (Roche Diagnostics, Indianapolis, Indiana) to determine the total plasma protein, albumin, calcium, phosphorus, glucose, blood urea nitrogen, creatinine, total bilirubin, creatine kinase, aspartate aminotransferase, alkaline phosphatase, globulin, albumin, gamma glutamyl transferase, magnesium, and plasma electrolytes (sodium, potassium and chloride). Activities of glutamate dehydrogenase and sorbitol dehydrogenase, were analysed using a Roche-Hitachi 717 analyzer (Roche Diagnostics, Indianapolis, Indiana).

Statistical Analyses

Data were analyzed using a GLM procedure (SAS 2002), with linear models fitting the effect of phenological stage. Nutritional composition of stems vs. leaves and offered vs. refusals were determined using the Student's t-test. Significance was set at ($P<0.05$).

RESULTS AND DISCUSSION

Crude protein values were higher in the rosette stage than in the bolting and flowering stages ($P<0.05$) (Table 1). No difference in crude protein content was found between bolting and flowering stages. Thomsen and co-workers (1989) reported a slightly lower value (10-13 % CP) for crude protein in the rosette stage and similar values (11-13% CP) for the bolting stage. The higher protein values in the rosette stage in our study may be due to the fertilization treatment that was applied. Values for acid detergent fiver (ADF) were higher ($P<0.05$) in the bolting and flowering stages than in the rosette stage. The values for ADF in the rosette stage (21.7%) were slightly lower than those reported by Thomsen et al. (1989) who reported 26-28%. In their study samples for the rosette stage were collected in late bolting stage right before bolting. Values for ADF in the bolting stage were similar to those reported by (Thomsen et al 1989) who reported 28-32% ADF. Neutral detergent fiber (NDF) content was higher ($P<0.05$) in the flowered than in the bolting and rosette stages and higher ($P<0.05$) in the bolting than in the rosette stage. Dry matter content was higher ($P<0.05$) in the flowering stage than in the rosette and bolting stage. No differences in dry matter content were found between rosette and bolting stages. Ash content was higher ($P<0.05$) in the rosette stage than in the bolting and flowering stage. No differences in ash content were found between the bolting and flowering stages. Gas production was higher ($P<0.05$) in the rosette stage than in the flowering stage and higher in the bolting stage than in the flowering stage. Energy content was higher ($P<0.05$) in rosette than in the bolting stage and was also higher ($P<0.05$) in the bolting stage than in the flowered stage.

Table 1. Nutritional composition of whole plants of yellow starthistle in three phenological stages, rosette, bolting and flowering. LSMEANS ± SD

		Phenological stage		
Nutritional characteristic	Alfalfa hay	Rosette	Bolting	Flowering
Dry matter (%)	91.8%[1]	93.3 ± 0.3^{a}	93.6 ± 0.2^{a}	94.3 ± 0.3^{b}
Organic matter (%)	89.8%[1]	85.5 ± 1.2^{a}	89.7 ± 0.7^{b}	90.4 ± 1.0^{b}
Ash (%)	10.8%[2]	14.5 ± 1.2^{a}	10.3 ± 0.7^{b}	9.6 ± 1.0^{b}
Crude protein (%)	21.7%[1]	17.3 ± 1.3^{a}	11.7 ± 0.7^{b}	9.6 ± 1.1^{b}
NDF (%)	41.5%[1]	27.5 ± 3.0^{a}	40.7 ± 1.7^{b}	47.7 ± 2.5^{c}
ADF (%)	32.5%[1]	21.7 ± 2.1^{a}	32.1± 1.2^{b}	34.5± 1.8^{b}
Gas 24 h.(ml/g DM)	41.86%[2]	47.6 ± 1.9^{a}	44.04 ± 1.1^{a}	39.2 ± 1.6^{b}
Energy Mcal/kg	2.5%[2]	2.3 ± 0.0^{a}	2.1 ± 0.0^{b}	1.9 ± 0.0^{c}
IVDOM	62.5%[1]	75.05 ± 2.0^{a}	66.33 ± 1.1^{b}	59.0 ± 1.7^{c}

^{a,b,c}Means on the same line with different superscript are significantly different $P< 0.05$.

[1] from (El-Hassan *et al.*, 2000).

[2] from (Getachew *et al.*, 2004).

Organic matter digestibility was higher ($P<0.05$) in rosette than in bolted and flowered stages and higher in bolted than in the flowering stage. In the bolting stage*in-vivo* digestibility ranged between 63.9 % and 87.9 % (mean 78.2 ± 1.5 %). The mean *in-vitro* digestibility in our study was 66.3 ± 1.1 %. This discrepancy between in vitro and in vivo may be due to the different methodologies used or the fact that for the gas test, rumen fluid from dairy cows with no previous exposure to yellow starthistle was used. It is possible that the secondary compounds present in yellow starthistle could cause a reduction in microbial activity. There is evidence of anti-microbial (Kuo-Hsing *et al.*, 1977; Bardon *et al.*, 2007), anti-protozoal (Yu *et al.*, 1994; Karamenderes *et al.*, 2006) and anti-fungal activity of sesquiterpene lactones (Panagouleas *et al.*, 2003; Bardon *et al.*, 2007). When mature leaves of *Centaurea maculosa* containing high concentrations of the sesquiterpene lactone cnicin were mixed with grass hay and fed to sheep, rumen gas production was significantly reduced (Olson and Kelsey, 1977). The potential effects of the secondary compounds in yellow starthistle in rumen microbial activity needs further research.

Table 2. Nutritional composition and in-vitro digestibility of leaves and stems of yellow starthistle in the bolting stage. LSMEANS ± SD

Nutritional characteristic	Leaves	Stems	Significance (P<0.05)
Dry matter (%)	92.8 ± 0.4	93.4 ± 0.8	NS
Organic matter (%)	84.7 ± 0.6[a]	91.5 ± 0.6 [b]	*
Ash (%)	15.4 ± 0.6[a]	8.5 ± 0.6 [b]	*
Crude protein (%)	16.2 ± 1.9[a]	10.4 ± 1.3 [b]	*
NDF (%)	25.9 ± 1.2[a]	36.8 ± 3.0 [b]	*
ADF (%)	22.6 ± 1.8[a]	34.6 ± 1.2 [b]	*
Gas 24 h (ml/g DM).	44.4 ± 1.5	42.3 ± 1.9	NS
Energy Mcal/kg	2.2 ± 0.1	2.1 ± 0.1	NS
IVDOM	71.6 ± 1.4 [a]	62.72 ± 2.2 [b]	*

Leaves of yellow starthistle in the bolting stage had higher nutritional value than stems as shown by higher crude protein content (P<0.05) and OM digestibility (P<0.05). Comparisons between leaves and stems are summarized in Table 2. Leaves also had higher (P<0.05) ash content than stems. Leaves also had lower fiber content (P<0.05) (NDF and ADF) and lower dry matter content (P<0.05) than stems. Consequently, digestibility of leaves was higher than that of the stems (71.6 ± 1.4 vs. 62.7 ± 2.2). The *in-vivo* digestibility of yellow starthistle in the bolting stage was 78.3 ± 6.2 %. These values are similar to those published by Callihan and co-workers (1995) who reported an *in-situ* dry matter disappearance of leaf samples ranging from 67 to 80% for prebolted and bolted starthistle. Despite the differences in crude protein, fiber and digestibility there were no differences between gas production and energy content between leaves and stems of yellow starthistle in the bolting stage.

The nutritional composition of offered and refusals showed significant differences. The refusals had lower protein (P<0.05), gas production (P<0.05) and energy content (P<0.05) and *in-vitro* digestibility (P<0.05) than the offerings. Refusals had also higher ADF content (P<0.05) than the offerings. The nutritional composition of offerings and refusals is summarized in Table 1.3. When fed yellow starthistle free choice, sheep were able to select the leaves that had higher nutritional quality and leave the stems that were less digestible and had lower crude protein content. During the feeding trials it was observed that the sheep would consume the leaves and the youngest part of the stems leaving out the woodiest part of the stems. Animals gained and average of 1.2 kg from beginning to the end of the feeding trial (total of 20 days).

Despite its good nutritional quality, yellow starthistle contains sesquiterpene lactones that could have anti-nutritional properties and affect the efficacy of sheep grazing as a control method. The anti-quality effects of sesquiterpene lactones in sheep range from mild toxicity to lethal. *Helenium microcephalum* is toxic to cattle sheep and goats due to the sesquiterpene lactones helenalin (Witzel *et al.,* 1976) and mexicanin-E (Kim, 1980) present in this species. Hymenovin, a sesquiterpene lactone found in*Hymenoxys richardsonii var. floribunda,Baileya multiradiata* (Ivie *et al.,* 1976), and *Dougaldia hoopesii* (Ivie *et al.*, 1976; Hill *et al.*, 1977) is also toxic to sheep. Vomiting disease of sheep in South Africa is due to Geigeris species which contains sesquiterpene lactones (Herz, 1978; Vigelzang *et al.*, 1978; Bohlman *et al.,* 1982). Hymenoxon, an epimeric form of Hymenovin, is found in *Helenium hoopesii*and causes "spewing sickness" in sheep (Hesker, 1982). When *Hymenoxis odorata,* containing hymenoxon, is added to the diet of sheep, voluntary intake decreases (Calhoun *et al.*, 1981).

Table 3. Nutritional composition and in-vitro digestibility of offered and refused yellow starthistle in the bolting stage. LSMEANS ± SD

Nutritional characteristic	Offered	Refusals	Significance (P<0.05)
Dry matter (%)	93.9 ± 0.2	94.5 ± 0.2	NS
Organic matter (%)	90.1± 2.0	90.6 ± 0.3	NS
Ash (%)	9.9 ± 2.0	9.4 ± 0.3	NS
Crude protein (%)	9.7 ± 0.3	6.7 ± 0.3	*
NDF (%)	47.0 ± 1.0	50.6 ± 1.0	NS
ADF (%)	35.1± 0.8	38.9 ± 0.7	*
Gas 24 h (ml/g DM).	46.2 ± 0.5	35.8 ± 0.6	*
Energy Mcal/kg	2.2 ± 0.0	1.8 ± 0.0	*
IVDOM	66.75 ± 1.4	55.80 ± 0.8	*

Our results show no evidence of anti-nutritional effects of bolted yellow starthistle for sheep. In our study, sheep were fed bolting yellow starthistle for a total of 65 days (May 15-July 19th) with no apparent signs of disease or increase in liver enzymes above their normal range for a healthy animal. All the liver enzymes measured were within normal range. Mean serum concentration of AP before ingestion of yellow starthistle was 112.1 ± 18.7 and after ingestion of yellow starthistle was 67.0 ± 16.2. Mean serum concentration of aspartate aminotransferase (AST) before ingestion of yellow starthistle was 67.4 ± 10.1 and after was 94.0 ± 71.9. Mean serum concentration of gamma glutamyl transferase (GGT) before ingestion of yellow starthistle was 54.5 ± 16.2 and after ingestion of yellow starthistle was 48.4 ± 13.8. Mean serum concentration of SDH before ingestion of yellow starthistle was 34.5 ± 6.2 and after ingestion of yellow starthistle was 43.5 ± 10.0. Mean serum concentration of glutamate dehydrogenase (GDH) before ingestion of yellow starthistle was 4.5 ± 2.2 and after ingestion of yellow starthistle was 6.6 ± 4.1. Serum concentrations of metabolites measured before and after ingestion of yellow starthistle were also within normal range. Serum concentrations of liver enzymes and metabolites before and after feeding yellow starthistle are summarized in table 4. Cordy (1954) reported a total absence of clinical signs after feeding four ewes and 2 lambs a total of 1200 lbs of 100% yellow starthistle over a period of 6 weeks. However, lambs fed bitterweed (*Hymenoxys odorata*) containing 0.68% hymenoxon had elevated levels of creatinine, urea nitrogen, GGT and AST (Calhoun *et al.*, 1989). The lack of an effect in either intake or the health consuming bolted yellow starthistle could be due to different reasons a) the sesquiterpene lactones present in bolted yellow starthistle do not have anti-quality properties as other sesquiterpene lactones or they are present at low concentrations to exert such effects, or b) that sheep have evolved strategies to cope with the sesquiterpene lactones present in yellow starthistle.

There are several strategies by which large herbivores minimize the anti-nutritional effects of secondary compounds ranging from changes in feeding behavior (i.e. diet selection) to the existence of detoxification mechanisms (Dearing *et al.*, 2005). Sheep are not avoiding the ingestion of the sesquiterpene lactones present in bolting yellow starthistle so these compounds must be either metabolized or excreted which would suggest the existence of detoxification or biotransformation mechanisms. Further research is needed to elucidate how sheep cope with the sesquiterpene lactones present in yellow starthistle. The etiology of equine encephalomalacia suggests that the toxic effects of yellow starthistle are exerted through ingestion over prolonged periods of time (Cordy, 1954). Horses show clinical signs of the disease after having consumed between 86-200% of their body weight in fresh plant material in feeding periods ranging from 26 to 81 days (average 54) (Cordy, 1978). Thus, it is possible that our experimental sheep did not ingest enough toxic compounds to significantly challenge or overcome the detoxification mechanisms. To be effective, grazing must be repeated for at least 3 consecutive years due to the survivability of seeds in the soil (DiTomaso *et al.*, 2000). Therefore, long-term studies on the effect of repeated and prolonged consumption of yellow starthistle in ruminants are needed.

Table 4. Serum concentrations of liver enzymes and metabolites before and after consumption of bolting yellow starthistle. Mean ± SD. AP= alkaline phosphatase, AST= aspartate amino transferase, GGT=gamma glutaryl transferase, SDH= sorbitol dehydrogenase, GDH= glutamate dehydrogenase, CK= creatine kinase

Parameters	Before YST	After YST	Reference
AP (iu/l)	71.7 ± 16.2	115.5 ± 19.9	50-300
AST(iu/l)	64.3.0 ± 11.3	63.5 ± 7.81	75-339
GGT (iu/l)	47.5 ± 16.23	51.0 ± 17.4	40-94
SDH (iu/l)	41.0 ± 4.5	33.0 ± 6.3	18-77
GLDH (iu/l)	6.2 ± 4.1	3.5 ± 1.1	2-10
Na (mmol/l)	144.3 ± 7.3	148.3 ± 3.4	140-150
K (mmol/l)	4.4 ± 0.7	4.9 ± 0.4	4.5-6.0
P (mg/dl)	5.6 ± 1.4	6.8 ± 0.6	3.8-8.7
Cl (mmol/l)	101.3 ± 7.1	107.7 ± 3.6	98-102
Calcium (mg/dl)	10.7± 0.2	11.3 ± 0.4	9.2-11.6
CK (iu/l)	109.3 ± 27.9	92.8 ± 21.5	100-547
Urea Nitrogen (mg/dl)	21.8 ± 4.7	11.2 ± 2.3	19-37
Glucose (mg/dl)	86.2 ± 12.7	62.7 ± 3.6	48-80
Creatinine (mg/dl)	0.8 ± 0.2	0.9 ± 0.1	0.8-1.3
Bilirubin (mg/dl)	0.2 ± 0.1	0.1 ± 0.0	0.1-0.2
Total Protein (mmol/l)	8.0 ± 0.5	7.3 ± 0.5	6.6-8.6
Total CO^2 (mmol/l)	18.2 ± 3.3	24.3 ± 1.6	20-26
Albumin (g/dl)	3.1 ± 0.2	3.0 ± 0.2	2.7-3.7
Globulin (g/dl)	4.9 ± 0.6	4.4 ± 0.5	2.8-5.4

Yellow starthistle is good forage for sheep, highly digestible and with adequate levels of protein and energy and low levels of fiber in all of its phenological stages. As yellow starthistle matures, its nutritional quality decreases as fiber content (NDF and ADF) increases and crude protein and energy content decreases. In the bolting stage its high nutritional value and lack of anti-nutritional factors makes yellow starthistle very palatable to sheep. Yellow starthistle in the bolting stage would provide enough ME and crude protein for a 60-70 kg ewe in maintenance (NRC 1985). Dry matter intake of bolting yellow starthistle ranged from

796.8 g/d (1.9 % BW)to 989.5 g/d (2.2% BW) (mean 931 ± 2 g/d or 2.03 ±0.02 in g/kg BW) which is comparable to the consumption of a forage of moderate quality. Intake levels of yellow starthistle for sheep should be measured in range conditions where sheep would have other dietary choices. In the late bolting stages, it is expected that sheep would preferentially consume yellow starthistle because its higher nutritional quality and moisture content makes it more palatable than the annual grasses characteristic of the rangelands in the Central Valley of California (Hart *et al.*, 1932). It is in the late bolting stage, when starthistle is still palatable to sheep and defoliation causes a reduction in seed production (Thomsen, 1993) and consequently when grazing would be most efficient as a control method. Sheep can and will graze bolted yellow starthistle without adverse health effects and can be an effective control method for this weed.

REFERENCES

Akbar, S., Fries, D. S., and Malone, M.H. (1995) Effect of various pretreatments on the hypothermic activity of repin in naïve rats. *Journal of Ethnolopharmacology, 49*, 91-99.

AOAC (Association of Official Analytical Chemists). (1990) *Official Methods of Analysis* (15th, 1200 pp.) Edition ed., Arlington, VA: Association of Official Analytical Chemists.

Bardon, A., S. Borkosky, M. I. Ybarra, S., Montanaro, S., and Cartagena E. (2007) Bioactive plants from Argentina and Bolivia. *Fitoterapia, 78*, 227-231.

Barry, S. (1995) *Cattle fecal fax.* University of California, Cooperative Extension. October Report, Nov. 7.

Benefield, C. B., DiTomaso, J. M., Kyser, G. B., and Tschohl, A. (2001) Reproductive biology of yellow starthistle: maximizing late season control. *Weed Science, 49*, 83-90.

Bohlmann, F., Zdero, C., and Ahmed, M. (1982) New sesquiterpene lactones, geranyllinalol derivatives and other constituents from Geigeria species. *Phytochemistry, 21*, 1679-1691.

Buttery, R. G., Madoxx, D. M., Light, D. M., and Ling, L. C. (1986) Volatile components of yellow starthistle. *Journal of Agriculture Food and Chemistry, 34,* 786-788.

Calhoun, M. C., Ueckert, D. N., Livingston, C. W., and Baldwin, B. C. (1981) Effects of Bitterweed (Hymenoxys-Odorata) on Voluntary Feed-Intake and Serum Constituents of Sheep. *American Journal of Veterinary Research,* 42, 1713-1717.

Calhoun, M.C., Baldwin, B.C., Kuhlmann, S.W., and Kim, H.L. (1989) Experimental prevention of bitterweed (*Hymenoxys odorata*) poisoning of sheep. *American Journal of Veterinary Research, 50*, 1642-1646.

Callihan, R. H., Lass, L. W., Hunt, C. W., and Pritchard, G. (1995) Comparative digestibility of yellow starthistle (*Centaurea solstitialis*). *Western Society of Weed Science*, 36.

Cordy, D. R. (1954) Nigropallidal Encephalomalacia in Horses Associated with Ingestion of Yellow Star Thistle. *Journal of Neuropathology and Experimental Neurology, 13*, 330-342.

Cordy D. R. (1978) Centaurea species and equine nigropallidal encephalomalacia. In R. F. Keeler, Van Kampen, K. R., and James, L. F. (Eds), *Effects of poisonous plants on livestock* (pp. 327-336). New York, NY: Academic Press.

Dearing, M. D., Foley, W. J., and McLean, S. (2005) The influence of plant secondary metabolites on the nutritional ecology of herbivorous terrestrial vertebrates. *Annual Review of Ecology Evolution and Systematics, 36*, 169-189.

DiTomaso, J. M., Kyser, G. B., and Hastings, M. S. (1999) Prescribed burning for control of yellow starthistle (Centaurea solstitialis) and enhanced native plant diversity. *Weed Science, 47*, 233-242.

DiTomaso, J. M. (2000) Invasive weeds in rangelands: Species, impacts, and management. *Weed Science, 48*, 255-265.

DiTomaso, J.M., Kyser, G. B. OrloffS. B., and Enloe, S.F. (2000) Integrated approaches and control option considerations when developing a management strategy for yellow starthistle. *California Agriculture, 54*, 30-36.

DiTomaso, J. M., Kyser, G. B., and Pirosko, C. B. (2003) Effect of light and density on yellow starthistle (Centaurea solstitialis) root growth and soil moisture use. *Weed Science, 51*, 334-341.

El-Hassan, S. M., Kassi, A. L., NewboldC. J., and Wallace, R. J. (2000) Chemical composition and degradation characteristics of foliage of some African multipurpose trees. *Animal Feed Science and Technology, 86*, 27-37.

Foley, W.J., Iason, G. R., and McArthur, C. (1999) Role of plant secondary metabolites in the nutritional ecology of mammalian herbivores: How far have we come in 25 years?. Vth. International Symposium on the Nutrition of Herbivores.American Association of Animal Science. pp 130-209.

Freeland, W. J., and Janzen, D. H. (1974) Strategies in Herbivory by Mammals - Role of Plant Secondary Compounds. *American Naturalist, 108*, 269-289.

Getachew, G., Robinson, P. H.,DePeters, E. J., and Taylor, S. J. (2004) Relationships between chemical composition, dry matter degradation and in vitro gas production of several ruminant feeds. *Animal Feed Science and Technology, 111, 57-71.*

Hart, G. H., Guilbert, H. R., and Goss, H. (1932) *Seasonal changes in the chemical composition of range forage and their relation to the nutrition of animals*. Bulletin 543. Berkeley: University of California, Agricultural Experiment Station.

Herz, W. (1978) Sesquiterpene lactones from livestock poisons. In:R. F. Keeler, van Kampen, K. R., and James, L. F. (Eds), Effects of Poisonous Plants on Livestock (pp. 487-497). New York, NY: Academic Press.

Hesker, K. (1982) Orange Sneezeweed: "Beautiful Flower of Death". *Rangelands, 4*, 2.

Hill, D. W., Kim, H. L., Martin,C. L., and Camp, B. J. (1977) Identification of Hymenoxon in Baileya-Multiradiata and Helenium-Hoopesii. *Journal of Agricultural and Food Chemistry, 25*, 1304-1307.

Ivie, G. W., Witzel, D. A., Herz, W. R., Kannan, J. O., Norman D. D., Rushing J. H., Johnson, L. Rowe, D., and Veech, J. A. (1975) Hymenovin - Major Toxic Constituent of Western Bitterweed (*Hymenoxys-Odorata* Dc). *Journal of Agricultural and Food Chemistry, 23*, 841-845.

Ivie, G. W., Witzel, D.A., Herz, W., Sharma, R.P., and Johnson, A. E. (1976) Isolation of Hymenovin from *Hymenoxys-Richardsonii* (Pingue) and *Dugaldia-Hoopesii* (Orange Sneezeweed). *Journal of Agricultural and Food Chemistry, 24*, 681-682.

Karamenderes, C., Khan, S., Tekwani, B. L., Jacob, M. R., and Khan, I. A. (2006) Antiprotozoal and antimicrobial activities of Centaurea species growing in Turkey. *Pharmaceutical Biology, 44*, 534-539.

Kim, H. (1980) Toxicity of sesquiterpene lactones. *Research Communications in Chemical Pathology and Pharmacology*, *28*, 189-92.

Kuo-Hsiung, L., Ibuka, T., Rong-Yang, W., and Geissman, T. A. (1977) Structure-antimicrobial activity relationships among the sesquiterpene lactones and related compounds. *Phytochemistry*, *16*, 1177-1181.

Launchbaugh, K. L., Provenza, F. D., and Pfister, J. A. (2001) Herbivore response to anti-quality factors in forages. *Journal of Range Management*, *54*, 431-440.

Maddox, D. M. (1981) Introduction, Phenology, and Density of Yellow Starthistle in Coastal, Intercoastal, and Central Valley Situations in California. *Science and Education Administration Publications* ARR: 1-33.

Makkar, H. P. S., Bluemmel, M., and Becker, K. (1995) Formation of complexes between polyvinyl pyrrolidones or polyethylene glycols and tannins, and their implication in gas production and true digestibility *in vitro* techniques. *British Journal of Nutrition*, *73*, 897-913.

Mamedov, G. M. (1956) *Centaurea solstitialis* from Azerbaidzhan. *AptechnDelo, 5*, 57.

Menke, K. H., and Steingass, H. (1988) Estimation of the energetic feed value obtained from chemical analysis and *in vitro* gas production using rumen fluid. *Animal Research and Development*, *28*, 7-55.

Merrill, G. B., and Stevens, K. L. (1985) Sesquiterpene lactones from Centaurea solstitialis. *Phytochemistry*, *24*, 2013-2018.

NRC. (1985) *Nutrient requirements of sheep*. (6th edition pp. 2-25), Washington D.C.: National Academy Press.

Olson, B. E., and Kelsey, R. G. (1997) Effect of *Centaurea maculosa* on sheep rumen microbial activity and mass in vitro. *Journal of Chemical Ecology*, *23*, 1131-1144.

Panagouleas, C., Skaltsa, H., Lazari, D., Skaltsounis, A. L., and Sokovic, M. (2003) Antifungal activity of secondary metabolites of *Centaurea raphanina* ssp *mixta,* growing wild in Greece. *Pharmaceutical Biology*, *41*, 266-270.

Pitcairn M. J., Schoenig, S., Yacoub, R., and Gendron J. (2006) Yellow starthistle continues its spread in California. *California Agriculture, 60*, 83-90.

Provenza, F. D., and Balph, D. F. (1987) Diet Learning by Domestic Ruminants - Theory, Evidence and Practical Implications. *Applied Animal Behaviour Science*, *18*, 211-232.

Roche, B. F., Roche C. T., and Chapman, R. C. (1994) Impacts of Grassland Habitat on Yellow Starthistle (*Centaurea-Solsitialis* L) Invasion. *Northwest Science*, *68*, 86-96.

Sheley, R.L., and Larson, L. L. (1994) Observation: Comparative live-history of cheatgrass and yellow starthistle. *Journal of Range Management*, *47*, 450-456.

Statistical Analysis Systems Institute (SAS). 2002. *SAS User's Guide: Statistics*, Version 8.1. SAS Inst. Inc., Cary,

Stevens, K. L., Riopelle, R. J., and Wong, R. Y. (1990) Repin, a sesquiterpene lactone from *Acroptilon repens* possessing exceptional biological activity. *Journal of Natural Products*, *53*, 218-221.

Thomsen, C.D., Williams, W. A., George, M.R., McHenry, W.B., Bell, F.L., and Knight, R.S. (1989) Managing yellow starthistle on rangeland. *California Agriculture*, *43*, 4-7.

Thomsen, C.D., Williams, W.A., Vayssieres, M., Bell, F.L., and George, M.R. (1993) Controlled grazing on annual grassland decreases yellow starthistle. *California Agriculture*, *47*, 36-40.

Van Soest, P. J., Robertson J. B., and Lewis, B. A. (1991) Methods for dietary fiber, neutral detergent fiber, and non-starch polysaccharides in relation to animal nutrition. *Journal of Dairy Science*, *74*, 3583-3597.

Vogelzang, M. E., Vermeulen, N. M. J., Potgieter, D. J. J., and Strauss, H. F. (1978) Ivalin in *Geigeria aspera*. *Phytochemistry*, *17*, 2030-2031.

Witzel, D. A., IvieG. W., and Dollahite, J. W. (1976) Mammalian Toxicity of Helenalin, Toxic Principle of Helenium Microcephalum Dc (Smallhead Sneezeweed). *American Journal of Veterinary Research*, *37*, 859-861.

Yu, H. W., Wright, C. W., Cai, Y. ,Yang, S. L. Phillipson, J. D., Kirby, G. C., and Warhurst, D. C. (1994) Antiprotozoal Activities of *Centipeda minima*. *Phytotherapy Research*, *8*, 436-438.

In: Grazing Ecology: Vegetation and Soil Impact
Editor: José Ramón Arévalo
ISBN 978-1-62100-436-3

Chapter 7

SOIL MOISTURE PATTERNS AND FLUXES RESPONDING TO GRAZING INTENSITY IN SEMIARID INNER MONGOLIA GRASSLAND

***Ying Zhao*[1,2], *Stephan Peth*[2] *and Rainer Horn*[2]**

[1]College of Resources and Environmental Science,
Northwest Agriculture and Forestry University, Yangling,
Shaanxi, P.R. China
[2]Institute of Plant Nutrition and Soil Science,
Christian-Albrechts-University zu Kiel, Olshausenstr.
Kiel, Germany

ABSTRACT

Overgrazing has been regarded as a main cause for land degradation in Inner Mongolia grassland because of the increase in population and shift in the socio-economic system in recent years. However, given the vital importance for the production of livestock and the environmental changes, fundamental understanding on topsoil sensitivity to grazing is lacking. Based on MAGIM (Matter fluxes in grasslands of Inner Mongolia as influenced by stocking rate) project, intensive observations, analyses, and modeling of environmental changes responding to grazing in Inner Mongolia grassland were conducted. In this chapter, we concentrated on the spatio-temporal variation of soil properties (e.g., soil moisture) in general and modeling the effects of grazing management (i.e., five grazing plots including, ungrazed since 1979=UG79, ungrazed since 1999=UG99, moderately grazed=MG, continuously grazed=CG, and heavily grazed= HG) on water flux in particular. Results showed that grazing treatment reduced the topsoil moisture contents (0-6 cm) and homogenized soil moisture spatial patterns which were controlled by soil and plant properties, especially soil physical properties. Multivariate geostatistics further revealed scale-dependent correlation for these controlling parameters depending on treatment. The soil and plant properties strongly controlled the variation of soil moisture for UG99 at short-scale (45 m), and for CG and HG at long-scale (90 m), however, weakly controlled the variation of soil moisture for UG79 and MG. Therefore, we suggest that pasture management strongly modified soil moisture patterns, which should be considered in the hydrological models at multiple spatial scales. Based on soil pit scaled monitoring data, we further found seasonal change of the soil moisture was most significant in the topsoil (0-10 cm) and was less pronounced in the deeper soil. Soil water storage decreased with increasing grazing intensity, which is linked to the grazing-induced soil structure change. Moreover, we use HYDRUS-1D model to parameterize this grazing-induced shifts of soil pore function. The model was well verified with the measured soil moisture values. The modeling results showed no significant difference on water budget components between the two ungrazed sites and moderate grazing, while heavy grazing significantly decreased interception from 17 to 7 mm and transpiration from 121 to 74 mm, and increased evaporation from 88 to 128 mm. We conclude that intensive grazing in our studied area deteriorated soil functions and reduced plant available water, and consequently reduced grassland productivity and enhanced the risks for wind and water erosion.

Keywords: Grazing impact, HYDRUS-1D, MAGIM, semivariogram, soil conservation

INTRODUCTION

Rangelands occupy about 30–50% of the Earth's land area, and they supply more than 80% of the feed of the livestock in Asia and Africa, 25% in North and Central America and 50% in the rest of the world (World Resources Institute, 2000). Inner Mongolia grassland (North China, Figure 1), one key part of the temperate Eurasian steppe belt from Eastern Europe to Eastern Asia, accounts for the largest Chinese grassland with about 20% of Chinese grassland areas (400 Mha). Grazing, therefore, is a major local land use form historically. This land is considered to be a fertile pasture and particularly flourishing in Yuan dynasty (A.D. 1271-1368). However, in the past few decades, due to the increase in population and shift in the socio-economic system, it is subject to an increased grazing pressure that leads to the broad degradation and risk of severe soil erosion, nutrient depletion and desertification (Li *et al.*, 2000). For instance, in the 1950's and 1960's local farmers were forced to give up their nomadic way of life and to settle in small villages, hamlets or individual farms (Sneath, 1998). Together with a moderate increase in number of livestock this policy has increased grazing pressure around the newly established settlements and brought the large scale pastoral movements between seasonal pastures to an end. The new economic policy established in the 1980's, which allowed individuals to profit directly from increased meat or wool production, sharply fostered the pressure on the land resources and resulted in intensified land-use and large scale overgrazing (Butterbach-Bahl *et al.*, 2011). Striking evidence for this process is given by numerous sandstorms attacking, e.g., Beijing with an increasing frequency. Yet, processes and factors responsible for the grassland degradation are not fully understood, although there is no doubt that grazing management makes a major contribution.

Given the vital importance for the production of livestock and the environmental changes, it is crucial to have a thorough understanding of the mechanisms that maintain or change the ecosystem in response to the changes of land management. To develop sustainable management strategies, it is essential to investigate the processes emerging from pasture practices. Since many of the rangelands are located in arid or semi-arid areas, water is obviously one of the key variables determining the fate of ecosystem situated in such regions. Thus the understanding of the hydrologic processes is critically important (Sugita *et al.*, 2007). Especially, the identification and prediction of soil moisture spatio-temporal patterns at different scales are necessary to understand underlying processes controlling the hydrological response of a region or even catchment (Grayson *et al.,* 2002, Lin *et al.,* 2006).

Figure 1. Location of the sampling plots delineated as: UG79: ungrazed since 1979, UG99: ungrazed since 1999, WG: Moderately grazed only in winter time (MG in the text), CG: continuous grazed, and HG: heavily grazed. Each plot has an area of 105 m × 135 m. Sampling points were spaced at 15 m with a subgrid spacing of 5 m (the orthogonal boxes with dash line in WG and CG). Two larger geostatistical grids in WG and CG had an area of 300 m × 550 m with a grid spacing of 50 m and a subgrid spacing of 10 m (no data in large grids were used in this study).

It has been accepted that environmental variables that are used to evaluate an ecosystem, e.g., soil properties, are spatial dependent (Goovaerts, 1998). The spatial dependency is commonly characterized and quantified by geostatistical methods such as autocorrelation and variogram analysis. Furthermore, multivariate geostatistics can deal with multivariate spatial data to classify the correlation of regionalization variables and to define scale-dependent relationships (Goovaerts, 1992). This is needed to extrapolate or interpolate variables from different spatial scales when employing either upscaling or downscaling procedures (Western *et al.,* 1999). For instance, when matter fluxes on various scales are investigated, it could be very useful to upscale water flux with the aid of physically-based hydrological models from plot to catena and finally to a regional scale.

Land management is reported to influence soil hydraulic properties by altering soil and plant functions (Flerchinger *et al.,* 2003). In pasture areas, soil mechanical disturbances due to animal trampling, interlinked with hydrological changes, often have detrimental effects on soil properties. Particularly in the topsoil, soil deformation is characterized by a decrease of pore volume and an altered pore size distribution, which both affect water and air conductivity (Willat and Pullar, 1983), and soil water retention characteristics (Martinez and Zinck, 2004).

These disturbances of soil structure further decrease water infiltrability and increase surface runoff. This will not only cause soil erosion by water and nutrient depletion, but also decrease the water availability and further increase soil erosion by wind. Thus, it is essential to quantify and predict management effects on soil properties to assess their consequences on plant production and the environment. To our knowledge, considering the stress-dependent changes of the environment, an adequate account of water flow processes and its spatial variability based on hydrological models is still lacking due to the complexity of the soil system.

Especially, considering the amount of plant available water as one of the most limiting factors for sustainable grassland development, the effects of grazing on water-related mechanisms and water budgets is urgently needed to be understood in our studied area (Chen and Wang, 2000). Some studies have shown that grazing increases evaporation but decreases transpiration because of the increasing bare ground area and reducing amount of biomass. Therefore, the heavily grazed sites are more susceptible to drought, and prone to being degraded due to the reduction of available soil water (Snyman, 2005). However, to which extent grazing affects evapotranspiration (ET), and how far ET is partitioned into transpiration and evaporation remains unclear (Leenhardt *et al.,* 1995). Furthermore, although water interception by the canopy and plant residues might occupy large volumes of water in steppe ecosystems, it normally is neglected or mixed with soil evaporation.

In the light of the absence of interdisciplinary studies of rangelands in northeastern Asia, a project called MAGIM (Matter fluxes in grasslands of Inner Mongolia as influenced by stocking rate) has been funded by DFG (Deutsche Forschungsgemeinschaft) for 6 years (2004-2010; website: http://www.magim.net). The principal objective of MAGIM is to understand how grazing of steppe ecosystems feedbacks on water, nitrogen and carbon fluxes on site and regional scales (Butterbach-Bahl *et al.,* 2011). This comprehensive task requires research on various scales involving detailed laboratory and field experiments as well as concurrent model development and application.

This chapter summarizes the current achievement in understanding the various causative factors controlling spatial soil moisture patterns affected by land management as well as the application and prediction of a physical-based hydraulic model HYDRUS (Zhao *et al.*, 2010, 2011a, 2011b). A combined spatial sampling and soil pit scaled monitoring were conducted on five plots, ranging from heavily grazed to ungrazed steppe ecosystem. Our objectives were (i) to quantify the spatio-temporal variability of topsoil moisture under different grazing intensity plots, (ii) to ascertain the main factors controlling soil moisture patterns at different spatial scales, (iii) to characterize and compare the temporal and vertical distribution of soil water regime in different grazing plots, and (iv) to quantify the grazing effects on water flux using a Richards' based hydraulic model.

MATERIALS AND METHODS

Study Area

The long-term experiment was conducted on the Inner Mongolia Grassland Ecosystem Research Station (IMGERS; 43°37′50″N, 116°42′18″E) situated in the Xilin River catchment

(3650 km^2, Inner Mongolia, Northern China). The vegetation is characterized by the perennial rhizome grass *Leymus chinensis* and bunch grass *Stipa grandis*, respectively. The region is marked by a continental climate with cold winter and warm summer. In the last two decades, the annual mean air temperature was 0.7°C and precipitation was 343 mm, of which more than 85% fell during the growing period (May to September). Although the rainfall season occurs during the growing period, plant water uptake is normally limited by both seasonal and/or annual rainfall fluctuations and the strong evaporative rate (Chen and Wang, 2000). The potential evaporation reaches 1750 mm yr^{-1}, so the water deficit per year may be as large as 1400 mm.

Five representative plots (hectare scale) with different sheep grazing levels were sampled (Figure 1): (1) 24 ha area ungrazed since 1979 (UG79); (2) 35 ha area ungrazed since 1999 (UG99); (3) 40 ha area moderately grazed only in winter time, with 0.5 sheep units (ewe and lamb) ha^{-1} yr^{-1} (MG); (4) 250 ha area continuously grazed, with 1.2 sheep units ha^{-1} yr^{-1} (CG); and (5) 100 ha area heavily grazed, with 2.0 sheep units ha^{-1} yr^{-1} (HG). The plots were adjacent to each other, apart from HG, which was 2.5 km away but on similar soils and topography. Prior to 1979, all plots were slightly grazed. All plots except UG79 were moderately grazed from 1979 to 1999 and fenced in 1999.

Spatial Sampling

At the plot scale, all measurements in each five plot were randomly arranged on a rectangular sampling grid, consisting of 100 points with a grid spacing of 15 m and an enclosed subgrid spacing of 5 m (Figure 1). The exact position of each point was located with a Real Time Kinematic Differential GPS (RTK-GPS). Furthermore, a terrain elevation model was derived, and topographical variables such as slope, aspect, curvature, and upslope contribution area were computed using ArcGIS (version 9.1, ESRI, Redlands, CA, USA; Johnston et al., 2001). A wetness index, which is a surrogate for lateral subsurface flow process, was also derived. This index was defined as $\ln(\alpha/\tan\beta)$, where α is the upslope contribution area and β is the local slope angle (Gómez-Plaza *et al.,* 2001).

At each plot, volumetric soil water content (0-6 cm) was measured both weekly and after rainfall events exceeding 3 mm, for 3 years during the growing season, starting in 2004. These measurements used theta-probes (Type ML2x, Delta-T Devices, Cambridge, UK) that were calibrated for the soil and inserted into the surface at each point. In total, measurements were made at 50 different times.

Soil water content was compared to other properties that were anticipated to influence water transport and storage. From June to August 2004, 0-4 cm depth samples were taken for soil organic carbon (SOC) determined in duplicate by dry combustion on a Vario Max CNS elemental analyzer (Elementar Analysensysteme GmbH, Hanau, Germany), and soil texture with the pipette method (Steffens *et al.,* 2008). Intact soil cores, 56 mm diameter x 40 mm heigh were taken for measurement of bulk density (BD). Vegetation coverage was analyzed using a non-destructive method based on Braun-Blanquet (1964). Aboveground biomass was sampled on 0.25 m × 0.25 m plots by cutting plant at 10 mm height, including the standing dead material. In addition, the following measurements were made *in situ*. Shear strength (SS) was measured with a hand-held shear vane tester (Geonor H-60, Norway). Water repellency was estimated with the water drop penetration time (WDPT) test, by measuring the time taken for a 0.5 mm^3 drop of water to enter the soil. Hydraulic conductivity was measured using a Mini-disk Infiltrometer (Decagon devices, USA) at a suction value of 0.5 cm. The main mean characteristics (100 points) for each plot are described in Table 1.

Table 1. The main soil and vegetation properties of the different grazing intensity plots sampled: UG79: ungrazed since 1979, UG99: ungrazed since 1999, MG: moderately grazed only in winter time, CG: continuously grazed, and HG: heavily grazed

Parameters	UG79 (n=98)	UG99 (n=99)	MG (n=100)	CG (n=88)	HG (n=98)
Sand	48.6 (3.2) b	46.1 (4.9) b	47.1 (5.5) b	44.6 (5.6) a	68.1 (4.7) c
Silt	35.2 (2.5) b	37.4 (3.4) c	35.2 (4.1) b	36.6 (3.5) c	20.9 (3.2) a
Clay	16.3 (1.0) b	16.5 (1.8) b	17.8 (1.6) c	18.7 (2.2) c	11.0 (1.9) a
SOC	31.0 (5.5) a	25.5 (6.3) b	–	22.2 (3.7) c	17.0 (4.1) d
BD	0.9 (0.10) a	1.1 (0.12) b	–	1.2 (0.06) c	1.3 (0.08) d
SS	29.3 (8.3) a	32.2 (7.2) b	35.1 (5.1) c	36.6 (9.9) c	43.6 (10.9) d
WDPT	16.4 (5.1) a	12.1 (4.8) b	6.0 (3.7) c	5.2 (2.1) c	1.7 (1.1) d
K	69.9 (19.1) a	157.8 (14.7) c	114.3 (23.1) b	126.0 (11.0) b	148.5 (11.8) c
SWC (pF=1.8)$^{\alpha}$	42.5 (4.1) a	41.7 (2.8) a	42.0 (8.7) a	41.4 (5.9) a	46.5 (6.1) b
SWC (pF=4.2)$^{\alpha}$	8.0 (2.6) b	9.8 (0.7) c	8.7 (2.6) c	9.2 (2.0) c	6.1 (0.7) a
VC	76.4 (8.8) a	68.0 (7.7) b	69.3 (3.9) b	54.5 (10.7)c	69.3 (11.6) b
AGB	313.1 (78.3) b	404.6 (70.5) a	160.8 (16.0) c	142.6 (17.9) cd	131.3 (41.3) d

Values in parentheses denote the standard deviation. Different letters give significant differences (p = 0.01) with ANOVA analysis. –: no data available. Sand: sand content (%); Silt: silt content (%);Clay: clay content (%); SOC: soil organic carbon (g kg^{-1}); BD: bulk density (g cm^{-3}); SS: shear strength (kPa); WDPT: water drop penetration time (s); *K*: hydraulic conductivity (cm d^{-1}); SWC: soil water content (%); pF: log matric potential in hPa; VC: vegetation coverage (%); AGB: Aboveground biomass (g m^{-2}).

$^{\alpha}$ value was derived by the neural network prediction tool ROSETTA (Schaap *et al.*, 2001) based on data of soil texture and bulk density.

Statistical and Geostatistical Methods

A factor complicating the study of soil moisture variability is that spatial patterns at a given scale may change with the field mean wetness condition. To quantify how the importance of these factors varies with the seasonal variations, we classified soil moisture measurements into three conditions. That is, for each plot and sampling time, the mean soil water content of all 100 points was used to delineate a soil moisture class as wet (>20%), moist (10-20%) and dry (<10%), based on the water content at wilting point (pF = 4.2) for the dry cut-off and half of field capacity (pF = 1.8) for the wet cut-off (Table 1). Sampling dates with the same wetness class for each plot were grouped, and then the average water content value for each individual point for these sampling dates was determined.

All measured variables were analyzed with descriptive statistics and geostatistics. Prior to statistical analysis, data were tested for normal distribution with a Kolmogorov-Smirnov test.

By removing one or two outliers (m ± 3s method) and data transformations (Ln, Cosine and Tanget), a normal distribution was obtained for each variable. The effect of grazing intensity on analyzed parameters was tested with a one-way analysis of variance (ANOVA) using SPSS 13.0.1 for windows (SPSS Inc., Chicago). Tukey's HSD method was used as post hoc test for multiple comparison of means (Steffens *et al.,* 2008). To determine the influence of each measured environmental factor on topsoil moisture patterns, a Pearson's correlation analysis was used.

The spatial variability was analyzed by geostatistical methods, which deal with how variance and covariance depend on the distance between observations. A spatial structure is commonly modelled by fitting a variogram function to an empirical semivariogram $\gamma(h)$ (Eq. 1). The $\gamma(h)$ is a plot of half the squared difference between two observations (the semivariance) against their distance in space, averaged for a series of distance classes:

$$\gamma(h) = \frac{1}{2N(h)}\left\{\sum_{\alpha=1}^{N(h)}[z(x_\alpha + h) - z(x_\alpha)]^2\right\} \tag{1}$$

where $N(h)$ is the number of pairs separated by the lag distance h, $z(x_\alpha)$ and $z(x_\alpha+h)$ are values of the measured variable at spatial locations x_α and $x_\alpha+h$, respectively. A variogram function is defined by its family and by three parameters: the sill (maximum semivariance), the range (the maximum distance at which pairs of observations influence each other), and the nugget effect (the variance within sampling units). The model families most commonly used (e.g., spherical, exponential and Gaussian) assume that there is no spatial dependence for distances larger than the range. In addition, the proportion of structure variance (difference between sill and nugget) to sill was used to define the spatial dependence; the higher the proportion, the stronger the dependence.

Furthermore, the cross-semivariogram is used to describe the pairwise difference relationships between the response variable (i.e., soil moisture) and explanatory variables (i.e., enviromental variables):

$$\gamma_{i,j}(h) - \frac{1}{2N(h)}\left\{\sum_{\alpha=1}^{N(h)}[z_i(x_\alpha + h) - z_i(x_\alpha)]\cdot[z_j(x_\alpha + h) - z_j(x_\alpha)]\right\} \tag{2}$$

where i and j denote each of the two variables considered. For properties that are positively correlated in space, positive covariances are calculated that increase with increasing distance between the sampling points. Negative covariance values are found if variables are inversely correlated in space. Following Deutsch and Journel (1998), we rescaled data values to range from 0 to 1 in order to equally compare values with different units, and limit variograms to a maximum separation of 90 m, which is one half the maximum distance between measured points, and is considered the largest reliable sampling domain.

In addition to semivariogram analysis, the Linear Model of Coregionalization (LMC) which proposes a general framework in multivariate geostatistics was used (Goovaerts, 1992; Casa and Castrignanò, 2007). The LMC is a set of models in which all auto- and cross-semivariograms are described by linear combinations of the same set of basic structures as follows:

$$\gamma_{i,j}(h) = \sum_{k=0}^{q} \gamma_{i,j}^{k}(h) = \sum_{k=0}^{q} b_{i,j}^{k} g^{k}(h) \tag{3}$$

where $b_{i,j}^{k}$ is the partial sill for the *i,j*th semivariogram for structure *k*, while $g^{k}(h)$ represents the basic type of variogram model for structure *k*. The $g^{k}(h)$ must be conditionally negative definite, whereas the coregionalization matrix of the coefficients $b_{i,j}^{k}$ must be positive semi-definite (Goovaerts, 1992).

The LMC assumes that an observed phenomenon is regarded as the sum of several independent subphenomena acting at different characteristic scales. We thus fitted an empirical variogram with a nested variogram model (Eq. 3) composed of several basic variogram functions with different parameters (i.e., range and sill). Experimentally these basic structures can be detected only if variograms are modeled as nested functions (Buttafuoco *et al.*, 2005). In such a case, it is possible to decompose the variogram into several spatial components, which then disclose the relationships between the variables at different spatial scales. The type of variogram model used in Eq. 3 was based on the experimental semivariograms of the variables and knowledge of the main geological and anthropogenic factors.

The scale-dependent correlations between regionalized variables was evaluated by the structural correlation coefficients, $\rho_{i,j}^{k}$

$$\rho_{i,j}^{k} = \frac{b_{i,j}^{k}}{\sqrt{b_{i,i}^{k} b_{j,j}^{k}}} \tag{4}$$

which is calculated from the partial sill value, $b_{i,j}^{k}$, of the cross-semivariogram model between *i* and *j* and the two partial sill values, $b_{i,i}^{k}$ and $b_{j,j}^{k}$, of semivariogram models for *i* and *j*, respectively (Goovaerts, 1992).

Long-Term *In Situ* Field Monitoring

In the pit scale (square-metre scale), four representative sites (except for CG plot) with different grazing intensities were investigated. The degree of plant and litter cover decreased with increasing grazing. At each site three replicate profiles with a distance of 15 m were installed and connected to one solar powered automatic data-logger that recorded data at 30-min intervals. In each soil profile, soil moisture and temperature were measured from 2004 to 2009 using horizontally inserted Theta-probes (Type ML2x, Delta-T Devices, Cambridge, UK) at 5, 20, and 40 cm depths and Platinum ground temperature probes (Pt-100) at 2, 8, 20, 40, and 100 cm depths, respectively (Zhao *et al.*, 2010, 2011b). Theta-probes were calibrated for the site-specific soil using the gravimetric method. At the initial installation, undisturbed soil samples (cylinder 100 cm^3, n = 7) were taken from four layers of 4-8, 18-22, 30-34, and 40-44 cm for the determination of soil physical properties. Saturated hydraulic conductivity was measured with a falling head permeameter. Water retention functions were determined with a ceramic pressure plate assembly by stepwise desaturating initially saturated samples at equilibrium matric potentials of -1, -3, -6, -15, -30, and -1500 kPa. After oven-drying of those samples, soil bulk density was measured. Pore size distribution was calculated from the water retention data according to parallel capillary model (Kutilek and Nielsen, 1994).

In 2004, a micrometeorological station in UG79 site was installed to record the standard observations such as precipitation, net radiation, and air temperature. To determine root distribution, a soil root auger (Cobra TT motorhammer, Eijkelkamp, The Netherlands) was used to take root samples up to 100 cm soil depth and separated into five layers of 0-10, 10-20, 20-50, 50-70, and 70-100 cm (Gao *et al.*, 2008). Root length was determined with the line intersection method using an 8× magnifying glass. Root measurements were conducted monthly during the growing period of 2004-2006. During root measurement, vegetation parameters (e.g., vegetation coverage, leaf area index (LAI), and plant height) were recorded. The shoot biomass was cut to ground level, transferred into plastic bags and stored at 5°C. Leaves and stems were separated, and LAI was measured with the leaf-area meter LI-3050 (LI-COR, Nebraska, USA). LAI Gaps were linearly interpolated to daily intervals.

Numerical Modeling

Simulations of soil water, vapor, and heat flow were performed with HYDRUS-1D (Zhao *et al.*, 2010). HYDRUS-1D is a finite element model for simulating the one-dimensional movement of water, heat, and multiple solutes in variably saturated media. The program numerically solves the Richards' equation for saturated and unsaturated water flow and Fickian-based advection dispersion equations for heat and solute transport (Šimůnek *et al.*, 1998). Variably saturated water flow above zero temperature is described by the modified Richards' equation (e.g., Hansson *et al.*, 2004):

$$\frac{\partial\theta}{\partial t} = \frac{\partial}{\partial z}\left[K_{\mathrm{Lh}}\frac{\partial h}{\partial z} + K_{\mathrm{Lh}} + K_{\mathrm{LT}}\frac{\partial T}{\partial z} + K_{\mathrm{vh}}\frac{\partial h}{\partial z} + K_{\mathrm{vT}}\frac{\partial T}{\partial z}\right] - S \qquad (5)$$

where θ [$L^3\,L^{-3}$] is the total volumetric water content (= $\theta_w + \theta_v$, θ_w and θ_v are the volumetric liquid water content and water vapor content, respectively), t is time, z is the spatial coordinate positive upward, h is the pressure head, T is the temperature, and S is a sink/source term usually accounting for root water uptake. Water flow in Eq. 5 is assumed to be caused by five different processes with corresponding hydraulic conductivities (given below in parentheses). That is, on the right-hand side, liquid water flows due to gradient in pressure head (K_{Lh}, [$L\,T^{-1}$]), gravity, and temperature (K_{LT}, [$L^2\,T^{-1}\,K^{-1}$]), and water vapor flows due to gradient in pressure head (K_{vh}, [$L\,T^{-1}$]) and temperature gradients (K_{vT}, [$L^2\,T^{-1}\,K^{-1}$]), respectively. Soil hydraulic properties are described by the following expressions (van Genuchten, 1980):

$$S_e(h) = \frac{\theta_w(h) - \theta_r}{\theta_s - \theta_r} = \frac{1}{(1 + |\alpha h|^n)^m} \qquad (6)$$

$$K_{\mathrm{Lh}}(h) = K_s S_e^l [1-(1-S_e^{1/m})^m]^2 \quad (7)$$

where S_e is the effective saturation, θ_s and θ_r are the saturated and residual water contents (L^3 L^{-3}), respectively; the symbols α (L^{-1}), n, and m = 1-1/n are empirical shape parameters, and the inverse of α is often referred to as the air entry value or bubbling pressure; K_s is the saturated hydraulic conductivity (L T^{-1}), and l is a pore connectivity parameter which normally is set to 0.5.

The potential evapotranspiration was estimated from the FAO Penman-Monteith equation (Allen *et al.,* 1998). The potential evaporation and transpiration was partitioned according to Beer's law as a function of cover area index of material covering the soil (Ritchie, 1972). Root water uptake was simulated using the model of Feddes *et al.* (1978). To reflect dynamics of plant water uptake, we employed a root growth model to describe the root distribution with time. It is important because soil water transport models combined with plant growth models enable a reasonable partitioning of calculated *ET* into actual evaporation and transpiration. In general, the maximum rooting depth was considered to be 100 cm, with the greatest root density in the upper 30 cm. The rooting depth increased linearly from 0 cm at the beginning of simulation to a maximum value at the date of "full cover" or harvest. According to the site-specific root measurements, the differences of root distribution between sites (e.g., root length density) were accounted for.

Empirically, the simulations of water and heat fluxes were subjected to various uncertainties resulting from parameterization of soil hydraulic characteristics and from the estimation of soil evaporation (Loos *et al.,* 2007). To analyze the effect of uncertainty in soil hydraulic parameters on the model output, three parameterization approaches were compared. Simulation 1 was conducted as a direct approach, using laboratory-derived hydraulic parameters (LDP) fitted by RETC code (van Genuchten *et al.,* 1991). In simulation 2, hydraulic parameters were derived by pedotransfer functions (PTFs) via the neural network (NN) prediction tool ROSETTA (Schaap *et al.,* 2001) based on soil texture and bulk density values. In simulation 3, an inverse approach (Inverse) was used to estimate hydraulic parameters employing a Levenberg–Marquardt parameter optimization algorithm in HYDRUS-1D. The hydraulic functions described in Eqs. 6 and 7 may contain up to six unknown parameters (θ_r, θ_s, α, n, K_s, and l). Hence, our four-layer soil profile (defined by soil horizons) may have as many as 24 unknown variables. To improve the optimization efficiency, the parameters α, and n were simultaneously optimized in all layers with other parameters (i.e., θ_r, θ_s, K_s, and l) constant using time series of observations of θ for each depth.

From phenological data at IMGERS, the growing season starts around late April and ends around early October. Thus we assigned the growing period, i.e., simulation time, to the period from 1 May to 30 September (in total of 153 days). The simulated soil profile was 100 cm deep with observation nodes located at 2, 5, 20, 40, and 100 cm. The model was calibrated with the measured data in 2004. For validation purposes, the model was applied to an additional 2 years of data (2005 and 2006) using the calibrated parameters from year 2004. Optimizing efficiency, as well as model efficiency was evaluated by the root mean square errors (RMSE):

$$RMSE = \sqrt{\frac{1}{N}\sum_{i=1}^{N}(P_i - O_i)^2} \tag{8}$$

where N is the number of observations, and P_i and O_i are the simulated and measured profile soil water values, respectively.

RESULTS AND DISCUSSION

Spatio-Temporal Variations of Topsoil Moisture

The ANOVA multiple comparison procedure revealed that the average values for soil organic carbon, bulk density, vegetation coverage and biomass are significantly different among different grazing intensity plots (Table 1). The heavily grazed plot (HG) had the least carbon and the greatest bulk density. In contrast, the plot protected from grazing for 25 yr (UG79) had the most carbon and the smallest bulk density. There exist consistent and significant relationships between grazing intensity and many of the measured properties, thus providing valuable field plots for further investigation of soil moisture variability in the different plots.

Topsoil Moisture Characteristics Derived from Grid Sampling and Analyses

Table 2 lists the descriptive statistics of soil moisture for each plot in the different classified moisture conditions. In general, soil water content (SWC) was greatest in UG99, followed by UG79, MG and CG, and smallest in HG, suggesting that grazing reduced soil water values. SWC was always greater in UG99 than in UG79, which can be explained by the larger leaf area of shrubs increasing rainfall interception and the thicker litter layer reducing water infiltration in UG79. Compared to the two plots grazed at moderate intensity (MG and CG), SWC in UG79 was greater after long dryness but smaller under moist and wet conditions. This could be due to the pronounced accumulation of organic matter observed, which would impede evaporation under dry water status (Zhao *et al.*, 2011a) and inhibit water infiltration under moist and wet water statuses, as indicated by the WDPT (Table 1). The smallest SWC was always observed in HG where scarce vegetal cover could have increased water evaporation.

Spatial Variability of Topsoil Moisture

There was a weak to moderate spatial dependency of soil moisture varying from 13 to 75% (Table 3), according to the classification of spatial dependency by Cambardella *et al.* (1994). In most cases, the variograms of soil moisture were expressed well by a spherical or exponential model, as indicated by the high coefficients of determination (R^2).

Table 2. Descriptive statistics of soil moisture for three soil water statuses measured during 2004-2006

Treatment	Sampling soil water statuses	Min.	Max.	Mean	Variance	Skewness	Kurtosis
UG79	<10% (Dry)	6	10	8	0.89	0.37	-0.24
	10-20% (Moist)	12	17	14	1.13	-0.17	-0.10
	>20% (Wet)	21	28	24	1.14	0.50	-0.05
UG99	<10% (Dry)	6	13	9	1.16	0.65	0.45
	10-20% (Moist)	13	23	18	1.43	1.39	0.35
	>20% (Wet)	25	30	27	1.19	-0.38	0.33
MG	<10% (Dry)	5	9	7	0.68	1.58	0.36
	10-20% (Moist)	13	17	15	0.75	-0.20	-0.52
	>20% (Wet)	23	27	25	0.82	-0.38	-0.35
CG	<10% (Dry)	5	11	7	1.11	1.22	1.03
	10-20% (Moist)	13	19	15	1.28	0.29	0.75
	>20% (Wet)	22	28	25	1.16	0.50	0.15
HG	<10% (Dry)	3	8	5	0.93	-0.55	0.42
	10-20% (Moist)	8	17	12	1.60	0.57	0.41
	>20% (Wet)	17	26	21	1.91	-0.21	0.31

Min.: Minimum (%); Max.: Maximum (%).

Compared with different plots, CG and HG had the relatively larger sills for most water statuses. A similar trend was also observed for the proportion of spatial structure, indicating a stronger spatial dependency with increasing grazing intensity. The smallest sill and proportion were found for the MG plot, which could be due to the development of a crusted soil (Krümmelbein et al. 2006). In the two ungrazed plots, UG79 had greater spatial variability than UG99 for moist and wet water statuses, potentially due to greater plant heterogeneity with a spotty distribution of shrubs (*Caragana microphylla*) after 25 yr protection from grazing. The range showed an erratic fluctuation between 47 and 170 m and was no clear trend with increasing grazing intensity. The smaller range in HG compared to CG and UG99 could be due to more partial and continuous compaction.

Interactions between Soil Moisture and Its Controls on Different Spatial Scale

Spatial variability of soil properties is somewhat inherent in nature because of variations in soil parent materials and microclimate. However, some of the variability may be caused by grazing and management practices.

Table 3. Isotropic variogram parameters of soil moisture for three soil water statuses measured during 2004-2006 for different grazing intensities

Treatment	Sampling soil water statuses	Nugget *Co*	Sill *Co+Cs*	Range A_0(m)	Proportion *Cs/(Co+Cs)*	R^2	RSS	γ-Model
UG79	<10% (Dry)	0.60	1.00	108	0.40	0.88	0.02	Exp
	10-20% (Moist)	0.93	1.38	63	0.32	0.83	0.07	Exp
	>20% (Wet)	0.90	1.46	47	0.38	0.95	0.01	Exp
UG99	<10% (Dry)	1.19	1.83	125	0.35	0.89	0.05	Exp
	10-20% (Moist)	1.07	2.97	168	0.64	0.95	0.13	Sph
	>20% (Wet)	0.69	1.54	111	0.55	0.89	0.12	Sph
MG	<10% (Dry)	0.51	0.59	55	0.13	0.67	0.00	Exp
	10-20% (Moist)	0.45	0.71	73	0.37	0.71	0.01	Exp
	>20% (Wet)	0.58	0.72	105	0.19	0.65	0.02	Sph
CG	<10% (Dry)	0.60	1.92	145	0.69	0.90	0.22	Sph
	10-20% (Moist)	0.84	2.32	149	0.64	0.88	0.32	Sph
	>20% (Wet)	0.61	1.93	170	0.68	0.92	0.16	Sph
HG	<10% (Dry)	0.30	1.19	93	0.75	0.80	0.13	Sph
	10-20% (Moist)	1.22	3.13	87	0.61	0.68	1.10	Sph
	>20% (Wet)	1.04	4.32	96	0.75	0.85	1.60	Sph

Cs: structural variance; R^2: determination coefficient; RSS: reduced sum of squares; Exp: Exponential; Sph: Spherical.

Both geologic/pedologic and land use factors interacted with each other on spatial and temporal scales, are additionally affected by plant covering and topography, which, in turn, are further modified locally by erosion and deposition processes. Therefore, non-stationary condition and spatial heterogeneity may occur and correlation analysis is necessary.

Correlation Analysis

A correlation matrix between SWC and potentially associated properties for the different soil water statuses is shown in Table 4.

Table 4. Pearson's correlation matrix between soil moisture and all causative factors for three soil water statuses for different grazing intensities (*P*<0.01 **; *P*<0.05 *)

	Parameters	UG79			UG99			MG			CG			HG		
		D	M	W	D	M	W	D	M	W	D	M	W	D	M	W
Soil	Sand	-0.14	-0.08	-0.21*	-0.28**	-0.26**	-0.50**	0.07	0.17	-0.05	-0.50**	-0.44**	-0.47**	-0.73**	-0.73**	-0.83**
	Clay	0.16	0.09	0.19	0.25*	0.24*	0.55**	-0.07	-0.18	0.05	0.55**	0.49**	0.55**	0.62**	0.65**	0.65**
	SOC	0.27*	0.19	0.13	0.29**	0.23*	0.50**	–	–	–	0.41**	0.35**	0.39**	0.55**	0.52**	0.52**
	BD	-0.24*	-0.19	-0.12	-0.30**	-0.23*	-0.45**	–	–	–	-0.32**	-0.22*	-0.24*	-0.38**	-0.39**	-0.34**
	SS	0.42**	0.20*	0.10	0.15	0.11	0.27**	0.13	-0.08	-0.02	0.14	0	-0.05	-0.32**	-0.19	-0.21*
	Ln(WDPT)	0.33**	0.18	0.03	0.21*	0.07	0.19	-0.06	0.03	-0.06	0.05	0.06	0.09	-0.35**	-0.32**	-0.39**
	Ln(*K*)	-0.31**	-0.12	-0.07	-0.08	-0.07	-0.12	-0.09	-0.17	-0.14	0.01	-0.02	-0.04	0.13	0.00	0.14
Plant	VC	-0.01	-0.01	0.22*	-0.07	0.05	0	-0.12	-0.04	-0.11	-0.25*	-0.22*	-0.34**	0.26**	0.22*	0.23*
	AGB	0.04	0.03	0.08	0.10	0.26*	0.33**	0.21*	0.14	0.04	-0.25*	-0.26**	-0.30**	-0.22*	-0.22*	-0.33**
Topo	RE	0.02	-0.13	-0.20*	0.05	0.18	0.19	-0.26*	-0.19	-0.05	-0.25*	-0.23*	-0.08	0.21*	0.10	0.15
-graphy	Tanget(SLO)	-0.01	0.21*	0.32**	0.06	0.12	0.20*	0.03	-0.01	0.06	-0.08	-0.09	-0.14	0.09	0.18	0.11
	Cosine(ASP)	0.01	0.14	0.03	-0.18	-0.12	0.12	0.30**	0.13	0.00	0.37**	0.40**	0.30**	-0.13	-0.07	-0.09
	CUR	0.09	0.01	0	0.13	0.18	-0.08	0.18	-0.01	0.05	-0.05	-0.02	-0.01	-0.01	-0.06	0.00
	Tanget(CAR)	-0.09	-0.08	0.03	-0.19	-0.03	0.10	-0.02	0.13	-0.07	0.01	-0.10	-0.10	0.01	0.08	0.02
	WI	0.05	-0.14	-0.20*	-0.12	-0.08	-0.01	-0.06	0.11	-0.07	0.08	0.03	0.10	0.04	0.01	0.03

D: dry; M: Moist; W: wet; Sand: sand content (%); Clay: clay content (%); SOC: soil organic carbon (g kg^{-1}); BD: bulk density (g cm^{-3}); SS: shear strength (kPa); WDPT: water drop penetration time (s); *K*: hydraulic conductivity (cm d^{-1}); VC: vegetation coverage (%); AGB: above ground biomass (g m^{-2}); RE: relative elevation (m); SLO: slope (degree); ASP: aspect; CUR: curvature; CAR: upslope contribution area (m^2); WI: wetness index. –: no data available.

For most grazing intensities and water statuses, a significant relationship between SWC and associated soil properties (e.g., soil texture, SOC and BD) was found. For UG99, CG and HG, there was a significant correlation between either sand content, SOC or BD and SWC. A weak correlation in UG79 might be a result of different developed history (e.g., ungrazed for 25 yr). This confirmed that the heterogeneity of soil properties is one of the most dominant factors affecting soil moisture variability, as supported by many other studies that also found SWC to be correlated to soil texture and BD (e.g., Gómez-Plaza *et al.*, 2001). Compared with soil physical factors, other hydraulic (WDPT) and mechanical parameters (SS) only showed a weak influence on soil moisture variability. Unexpectedly, we even did not detect a clear correlation between SWC and hydraulic conductivity (K) since many studies envisioned soil moisture variability should be strongly controlled by variations in K (Hawley *et al.,* 1983; Hébrard *et al.,* 2006). It is well known that K depends strongly on pore size and continuity, wetting properties and antecedent SWC. Considerable temporal variability in all of these properties on K, combined with their different impacts, may have confounded any clear trend being observed.

Plant heterogeneity may be another important controlling source of soil moisture variability. Irrespective of wetness conditions, SWC was significantly correlated with vegetation coverage (VC) and aboveground biomass (AGB) in the continuously or heavily grazed plots (Table 4). However, in the ungrazed plots, correlations were only observed under a wet water status in UG79, and under moist and wet water statuses in UG99. Hawley *et al.* (1983) found that vegetation coverage was a major factor influencing soil moisture variability, and further noted that differences were often greater in wet conditions than in dry conditions. However, we did not find this trend. In the HG plot, the positive correlation between VC and SWC showed the shading effect of vegetation. In contrast, AGB was negatively correlated to SWC, which might imply the deleterious impact of heavy grazing on vegetation productivity (i.e., plant community succession of degradation by shortgrass).

With minor exceptions, neither surface factors, e.g., relative elevation, slope and aspect (local control) nor subsurface factors, e.g., curvature, upslope contributing area (CAR) and wetness index (WI) (unlocal control) exhibited a significant correlation with soil moisture for any soil water statuses (Table 4). The aspect was significant only for the CG plot, where the topography was relatively flat. The relative elevation was important at the grazed plots under dry conditions, while the slope was a key factor at the ungrazed plots under wet conditions. Under wet conditions, unlocal subsurface flow may mostly influence SWC (Grayson *et al.,* 1997), but this was not found for any investigated plots. Thus topography had only a slight effect on soil moisture distribution in our studied area.This may relate to the local topographical conditions. The experimental site is characterized by a smooth wide plain with relative height of 20-30 m and slopes <10° where no clear mounds, depressions, and therefore no significant differences in slope, aspect and curvature. The similar result was also found by Hébrard *et al.* (2006).

Multivariate Geostatistics

As shown in Table 4, SWC was influenced by several soil and plant parameters. However, single scale correlation analysis may disregard the effects of spatial scale. For this approach, a geostatistical analysis was performed. The semivariogram analysis was carried out only for elevation, sand content, clay content, SOC, BD, VC and AGB, since they showed the relatively strong correlations with SWC compared to other variables (Table 4). As soil water status was shown to have limited influence on the controlling factors (Zhao *et al.*, 2011a), all SWC data were averaged for each individual plot.

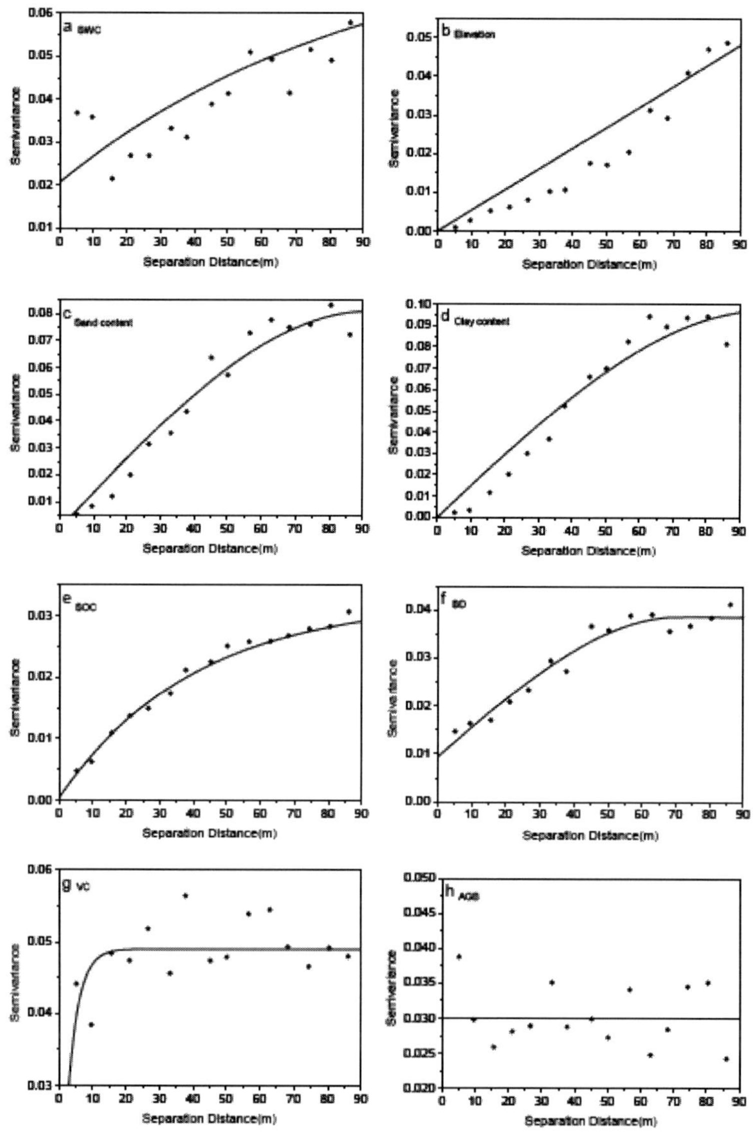

Figure 2. Semivariograms of (a) SWC, (b) Elevation, (c) Sand content (%), (d) Clay content (%), (e) SOC, (f) BD, (g) VC, and (h) AGB in UG99, with the model fitted. SWC: soil water content (%); SOC: soil organic carbon (g kg^{-1}); BD: bulk density (g cm^{-3}); VC: vegetation coverage (%); AGB: Aboveground biomass (g m^{-2}).

The isotropic semivariograms for the eight selected parameters are exemplarily shown for UG99 (Figure 2). Except for elevation and AGB, other parameters presented a moderate to strong spatial dependence, which can be fitted well by spherical or exponential models. The variation of elevation continuously increased with separation distances, as fitted by a linear model. In contrast, the variation of AGB showed the pure nugget effect, indicating a strong random variation. Each variogram was the basis for the mapping by a kriged interpolation scheme of corresponding parameters. The ordinary kriged maps revealed matching/ mismatching in spatial patterns between SWC and other environmental factors (Figure 3).

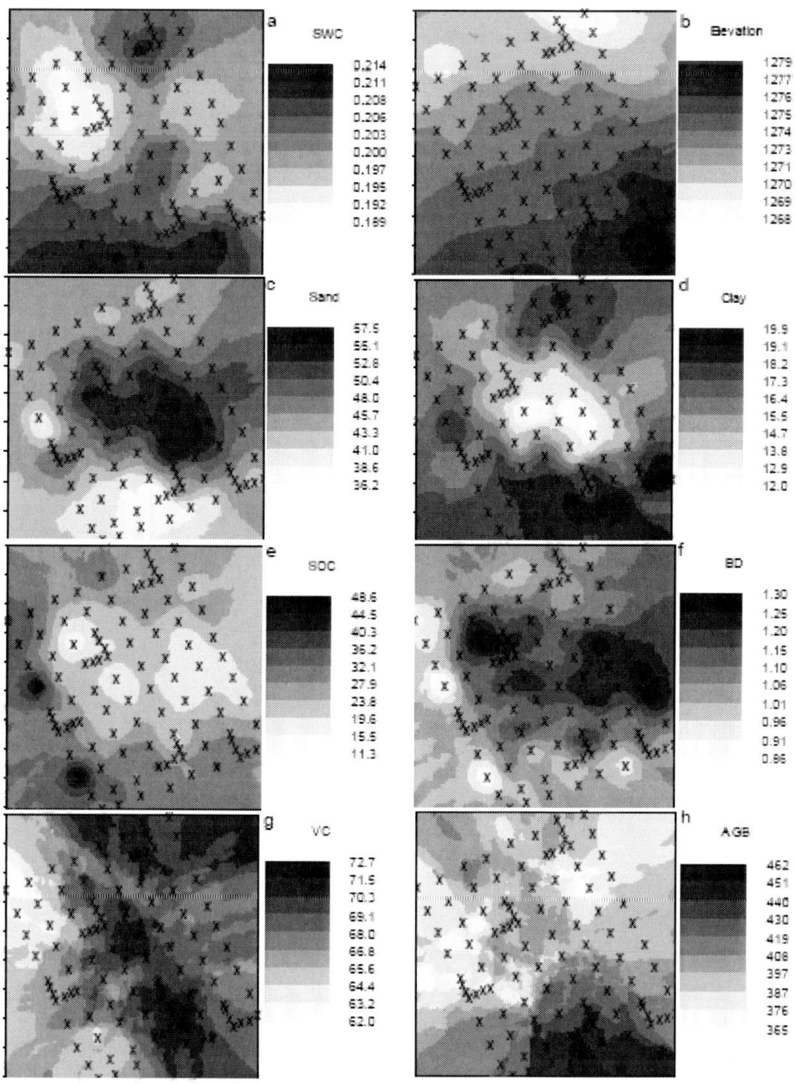

Figure 3. Ordinary kriged maps of (a) SWC, (b) Elevation, (c) Sand content, (d) Clay content, (e) SOC, (f) BD, (g) VC, and (h) AGB in UG99. SWC: soil water content (10^2%); SOC: soil organic carbon (g kg^{-1}); BD: bulk density (g cm^{-3}); VC: vegetation coverage (%); AGB: Aboveground biomass (g m^{-2}).

For instance, the central part showed lower values of SWC, clay content and SOC, but higher values of sand content and BD than that at the two edges. This was in line with the correlations (Table 4). Compared with soil variables, two vegetal variables, VC and AGB, showed a relatively patchy distribution with intermittent local highs and lows which did not clearly match the distribution of SWC.

Furthermore, a multivariate geostatistics was conducted to detect the interrelations among water-related variables at the multiple scales. To simplify the dataset, the coregionalization modeling was only carried out for SWC, sand content, BD and AGB. Since it is close correlation either between sand content and silt content, or between SOC and DB, we considered it is available and efficient to choose one from two. Similarly, AGB is chosen as a surrogate for vegetative variables. The LMC was fitted using three spatial structures: nugget effect (random; <5 m), a spherical model with a short-range (45 m) and a spherical model with a long-range (90 m) based on Eq. 3. The two ranges were observed in the experimental semivariogram (Figure 2) and cross-semivariogram (Figure 4), which was probably related to intrinsic properties of soil particle and aggregate distributions at the short-range scale and field size at the long-range scale.

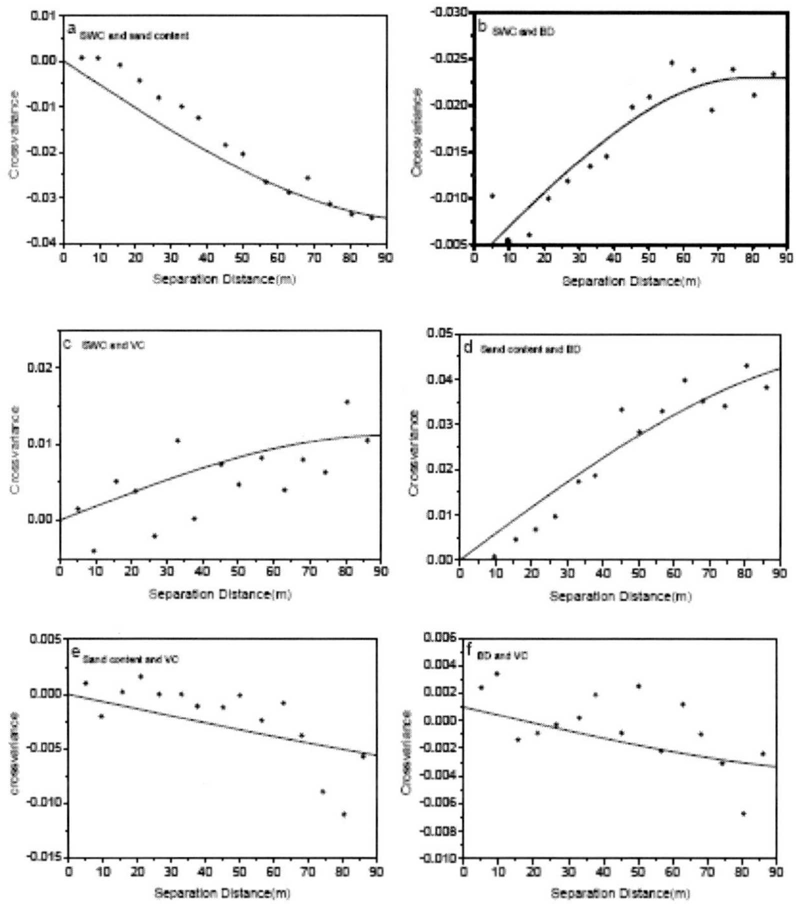

Figure 4. Cross-semivariograms of (a) SWC and sand content, (b) SWC and BD, (c) SWC and VC, (d) Sand content and BD, (e) Sand content and VC, and (f) BD and VC in UG99, with the linear model of coregionalization fitted. SWC: soil water content (%); BD: bulk density (g cm^{-3}); AGB: Aboveground biomass (g m^{-2}).

The random structural correlation coefficients were zero (data not shown), as no nugget effects were obtained from the cross-semivariogram analysis. The cross-semivariograms confirmed that SWC and its causative factors were not only statistically, but also spatially highly correlated (Figure 4), suggesting that soil moisture response to environmental heterogeneity is scale-dependent. From the structural correlation coefficients corresponding to the different scales (Table 5), we found that SWC had higher correlation coefficients with soil variables (i.e., sand content and BD) than vegetative variable (i.e., AGB) at both scales.

Table 5. Structural correlation coefficients between soil moisture and soil and plant properties at short- and long-range scales for different grazing intensities

Treatment		Short-scale				Long-scale			
		SWC	Sand	BD	AGB	SWC	Sand	BD	AGB
UG79	SWC	1.00				1.00			
	Sand	-0.68	1.00			-0.10	1.00		
	BD	0.82	-0.40	1.00		-0.89	0.38	1.00	
	AGB	-0.16	0.76	-0.15	1.00	0.00	-0.69	-0.41	1.00
UG99	SWC	1.00				1.00			
	Sand	-0.99	1.00			-0.57	1.00		
	BD	-1.00	0.99	1.00		-0.15	0.71	1.00	
	AGB	-0.68	0.76	0.66	1.00	0.74	-0.56	-0.65	1.00
MG	SWC	1.00				1.00			
	Sand	–	1.00			–	1.00		
	BD	–	–	1.00		–	–	1.00	
	AGB	–	–	–	1.00	–	–	–	1.00
CG	SWC	1.00				1.00			
	Sand	-1.00	1.00			-0.75	1.00		
	BD	-0.30	0.39	1.00		-0.93	0.88	1.00	
	AGB	-0.32	0.42	1.00	1.00	-0.57	0.53	0.38	1.00
HG	SWC	1.00				1.00			
	Sand	-0.89	1.00			-0.88	1.00		
	BD	0.63	-0.22	1.00		-0.97	0.73	1.00	
	AGB	0.55	-0.86	-0.31	1.00	-0.94	0.99	0.82	1.00

SWC: soil water content (%); Sand: sand content (%); BD: bulk density (g cm^{-3}); AGB: above ground biomass (g m^{-2}). –: no data available.

This may be explained by the high nugget variations of plant properties. The total spatial variation of soil moisture was mostly dominated by variations of soil variables within a short-range scale in both ungrazed plots UG79 and UG99, but a long-range scale in CG and HG. This difference could be due to grazing intensity as the regionalized variables had a more heterogeneous spatial distribution in the ungrazed plots compared with the grazed plots. We ascribed this to be the result of direct physical impacts of grazing sheep on topsoil, disturbance effects of grazing on vegetation, and potentially soil erosion (wind and water) due to the vulnerability of soil functions and reduction of vegetation coverage (Table 1). This is in agreement with Adler and Lauenroth (2000), who found heterogeneous, patchy distributions of vegetation and soil in ungrazed areas or following grazing exclusion compared to homogeneous spatial distributions in grazed areas. Specifically, the soil and plant properties strongly controlled the variation of soil moisture for UG99 at short-scale (45 m), and for CG and HG at long-scale (90 m), however, weakly controlled the variation of soil moisture for UG79 and MG.

Noted that some correlations changed between positive and negative when moving from one spatial scale to the other, e.g., BD in UG79 and HG, illustrating that SWC might be controlled by the same physical processes that operate at different spatial scales and consequently influenced by completely different ways. Therefore, such scale-dependent relations are critical to define the spatial scale of a simulation model when devising an ecological monitoring/sampling scheme.

Response of Soil Moisture to Grazing Effects

Spatial and temporal monitoring of soil thermal and moisture regimes are basic and essential tools for understanding additional emerging processes that degrade pastures. In arid and semiarid regions, vegetation cover could be strongly controlled by the water availability, especially considering the plant response to normal seasonality, to between-year irregularities and to extreme drought events (Casas and Ninot, 2007).

In Inner Mongolia grassland, the assessment of grazing impacts on soil water have been limited to a single growing season, shallow soil depths, and short-term grazing treatments (Li *et al.,* 2006). Identifying seasonal and annual variation in this area requires well documented analyses of soil water and temperature regimes, based on long-term experimental studies.

Four Hydric Periods

For the analysed three plots (UG79, MG and HG), soil moisture showed roughly similar seasonal dynamics with the regular yearly cycles throughout the investigated six years that spanned between June 2004 and September 2009 and experienced a large range in precipitation between 162 and 362 mm (Figure 5). The years 2004 and 2009 were more humid than 2005, 2006, 2007 and 2008. In general, we are able to segregate four hydric periods within each hydrological year (Zhao *et al.*, 2011b):

i. Growing period (Late April–August), when the soil water content remained relatively high (normally >10%), and the soil temperature is normally higher than 10ºC This phase is the key period for grass growth in the studied area, where high available water coincides with high temperature.
ii. Transitional period from summer to winter (September–October), when temperature and potential evapotranspiration decrease, and plants stop growth.

iii. Winter time (November–First March), when the temperature is below 0ºC, and soil water is mostly frozen, thus the liquid water content is less than 5%, and the matric potential is therefore extremely negative. The low precipitation and strong wind usually result in a shallow, but often unevenly distributed snow-cover.
iv. Transitional period from winter to summer (March–April), when temperature and potential evapotranspiration increase, and may result in an intense surface runoff from the frozen soil layer. Ice and snow melting recharges the water storage and initiates plant growth.

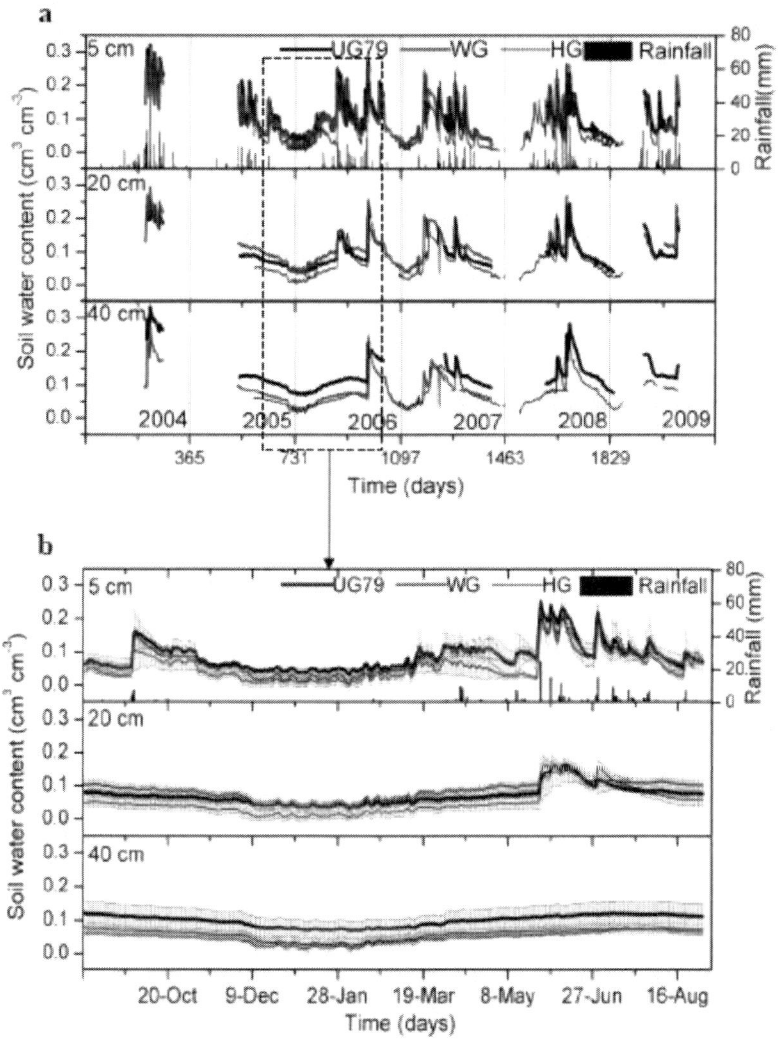

Figure 5. a, Daily changes of soil water contents (from 2004 to 2009) in the three investigated sites at 5, 20, and 40 cm soil depths; b, The daily changes of soil water contents in a hydrological year from September 2005 to August 2006. UG79: ungrazed since 1979, MG: moderately grazed only in winter time, and HG: heavily grazed.

Seasonal and Interannual Variability of Soil Moisture Affected by Grazing Intensity

During the whole monitoring period, soil water content (liquid) showed strong seasonal variations fluctuated with precipitation and temperature, ranging from 0.02 to 0.32 cm^3 cm^{-3} for UG79, 0.01 to 0.31 cm^3 cm^{-3} for MG, and 0.01 to 0.26 cm^3 cm^{-3} for HG (Figure 5a). The seasonal precipitation in 2006 provided a good example for analysing the dynamics of soil water in a single year (Figure 5b). Low standard deviation indicated a small spatial variability, and thus soil water varied slightly between different monitoring sites within each plot with relatively flat terrains. A major increase in soil water due to higher repeated precipitation events occurred primarily in June and July. However, these water dynamics were more pronounced in the topsoil (0-10 cm) compared to the subsoil, indicating a small water infiltration into deep soil.

Although temporal patterns of water content were not affected by grazing, the water redistribution in the profile was clearly affected (Figure 5). In general, the soil water content was always lowest in HG for all depths throughout the year. In the deeper soil layer (30-50 cm), soil moisture in UG79 remained relatively constant and significantly (p <0.05) higher than in the other two plots for all hydric seasons (Figure 6).

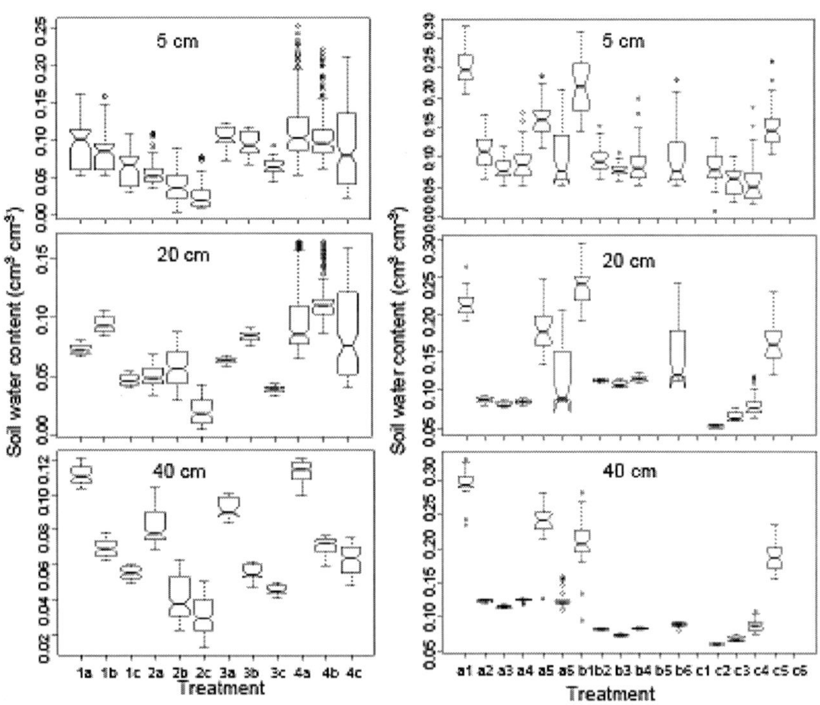

Figure 6. The box-plot showing the range of soil moisture for each grazing treatment from September 2005 to August 2006 with different hydric periods. 1: first hydric period, 2: second hydric period, 3: third hydric period, and 4: fourth hydric period (left side); and the ranges of soil moisture for different grazing treatments in August for 6-yrs periods from 2004 to 2009. 1: 2004y, 2: 2005y, 3: 2006y, 4: 2007y; 5: 2008y, 6: 2009y (right side). a: UG79, b: MG, and c: HG. UG79: ungrazed since 1979, MG: moderately grazed only in winter time, and HG: heavily grazed.

For the topsoil (0-10 cm), these differences were significant either only during the winter and winter transitional periods for all sites, or only between UG79 and HG for all hydric periods. However, in the intermediate-depth layers 10-20 cm, soil water contents were significantly greater in MG, compared with ungrazed site UG79, where more rainfall water appeared to percolate into deep soil layers.

Water Storage and Plant Available Water

Based on the *in situ* soil water measurements, soil water storage (from 0 to 50 cm) was calculated for the hydrological year 2005/2006 (Figure 7). Soil water storage was significantly greater in the ungrazed site compared to the grazed sites ($p < 0.05$), indicating that grazing reduced the water stored in soil, especially in the deep soil. Even under soil frozen conditions, it was still greater in the ungrazed site than in the grazed sites, suggesting that the volume of unfrozen water was also affected by the water content prior to soil freezing.

Figure 8 shows the dynamics of matric potential and plant available water at the three sites during the growing period. Water potential in the topsoil remained around the value matching with half of field capacity from June to July. At the beginning of the growing period, water potentials in HG remained around the plant wilting point in the topsoil, which resulted in earlier plant drought stresses. This may have resulted from lower water storage before freezing in October, 2005, and also from water loss above the frozen soil layer in spring melting time. The limited soil water replenishment after thawing in combination with plant water uptake during germination both could be responsible for the completion of the topsoil water storage (0-10 cm) in the HG site by the end of May.

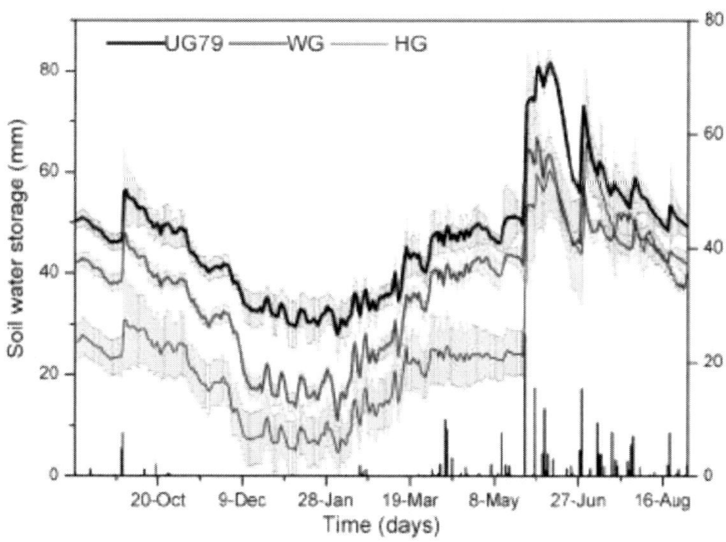

Figure 7. Daily changes of soil water storage from September 2005 to August 2006 in the three investigated sites. UG79: ungrazed since 1979, MG: moderately grazed only in winter time, and HG: heavily grazed.

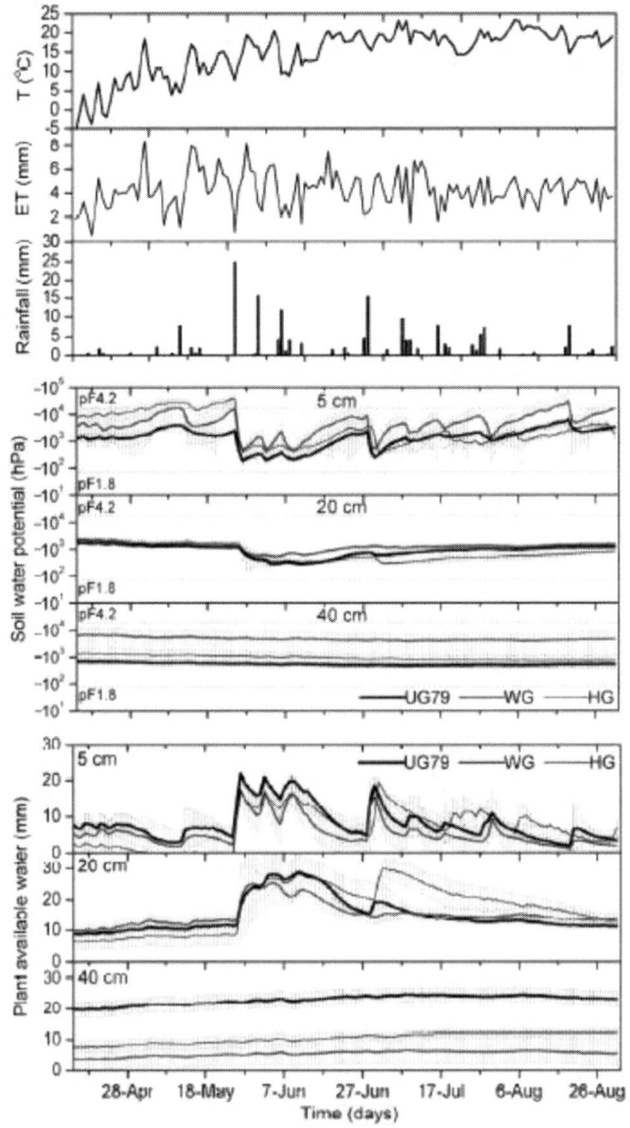

Figure 8. Soil water storage, plant available water, rainfall, evapotranspiration (ET) and air temperature (T) in the growing period of 2006. UG79: ungrazed since 1979, MG: moderately grazed only in winter time, and HG: heavily grazed.

However, the actual plant available water (0-50 cm) between the treatments varied throughout the year as it was initially greater in the ungrazed site than in the heavily grazed site during the first growing period (May-June), but smaller during the second growing period (June-August). At 30-50 cm depth, the amount of plant available water was significantly higher ($p < 0.05$) at UG79 than at HG and MG, but this site-specific difference was not clear in the other two depths (0–10 or 10-30 cm). Thereafter, the topsoil was occasionally recharged by rains, showing several peaks during the growing season. In contrast, soil water contents at 20 cm and 40 cm depths were gradually depleted from May to September with almost no signs of recharge except for occasional increases at the 20 cm depth following great rain events.

Hydrologic Processes and Model Predictions for Soil Moisture

Butterbach-Bahl *et al.* (2011) summarized current knowledge and recent experimental findings that intensive grazing clearly decreased soil aggregation and the amount of fresh, litter-like particulate organic matter (Kölbl *et al.,* 2011). As a consequence of weak aggregation in combination with animal trampling, soil organic matter mineralization is enhanced, whereas topsoil bulk densities were increased and infiltration rates, saturated hydraulic and air conductivities were decreased (Reszkowska *et al.,* 2011). The effects of soil structure disruption due to grazing was further amplified by the degradation of vegetation patches and resulted in a texture controlled wettability of the soil surface (Table 1; Kölbl *et al.,* 2011). These grazing induced changes within the study area accelerated the influence of soil and vegetation parameters on soil moisture and lowered soil water storage and plant available water during all seasons (Zhao *et al.,* 2011b).

Grazing-Induced Shifts in Hydraulic Properties

Pore size distribution curves for different grazing intensities are shown in Figure 9. The distinct shift in the curve shapes indicated a change in pore size at the grazed sites compared to the ungrazed site. In the topsoil (0-10 cm; normally assumed as the influence scope of animal hooves), animal trampling induced soil compaction, resulting in significantly reduced total porosity, particularly in HG site. In the subsoil, however, the total pore volume was the greatest in MG site. Near water saturation (i.e., matric potential >-10 hPa), the water retention curve of the ungrazed site decreased steeply in contrast to the two grazed sites, indicating a large fraction of macropores (>50 μm) at the undisturbed control site. Due to a higher trampling intensity in HG site, the macropores fraction of this site with the coarse-textured was much smaller than that of the fine-textured MG site. This change in pore size distribution also showed a noticeable effect on the hydraulic conductivity function. That is, consistent with the soil water characteristics, the predicted unsaturated conductivity $K(h)$ curves showed similar trends between treatments (data not shown, Zhao *et al.*, 2010).

Model Calibration and Validation

The previous differences of soil moisture and temperature between sites provide some suggests that fluxes of water and heat interact depending on the grazing intensity. Before modeling these coupled fluxes, model calibration was done for each site using site-specific boundary conditions including soil temperature, partitioned ET and interception. During the growing period of 2004 to 2006, LAI was significantly smaller in HG than in UG79 and MG, while the latter two sites were similar except that LAI in UG79 was significantly greater than in MG at the peak biomass time in 2004 (Zhao *et al.,* 2010).

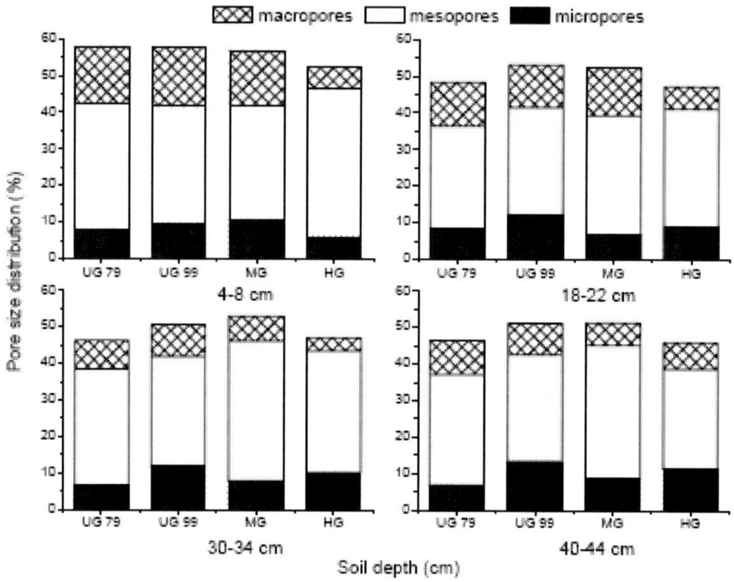

Figure 9. Pore size distribution for the four sites at four soil depths. Sparse: macropores size (>50 μm, pF <1.8), pF 1.8 = field capacity; White: mesopores size (0.2-50 μm, pF 1.8-4.2), available water capacity; and Black: micropores size (<0.2 μm, pF <4.2), pF 4.2 = permanent wilting point. UG79: ungrazed since 1979, UG99: ungrazed since 1999, MG: moderately grazed only in winter time, and HG: heavily grazed.

As a result, we predicted larger Tp/ET ratios for the ungrazed sites than for the grazed sites. Regarding ET partitioning, the estimation of potential transpiration based on detailed LAI measurements are essential for a reliable calculation of Tp. In the HYDRUS-1D model, Tp is determined by both potential transpiration and root water extraction.

The optimized parameters α, and n based on the inverse optimization were clearly different from the measured values (LDP) or statistical estimations (NN). There were large differences in $K(h)$ among the three parameterization approaches (Zhao *et al.*, 2010). The quality of the model results were tested by field observed water content data. The general trend for simulated water content, based on the different simulation approaches, was similar for all sites. An example for UG79 is shown (Figure 10). The Inverse and LDP provided good predictions, except that the LDP underestimated soil water content at 40 cm depth. However, the NN poorly predicted soil water content for all soil layers. Model accuracy was further assessed using RMSE. In each site, RMSE was smallest for the Inverse exhibiting the best goodness-of-fit, followed by LDP and NN (Table 6). This suggests that the optimized parameters are more accurate than the simulations via laboratory measured data and estimations based on PTFs.

The hydraulic parameters calibrated to the data for year 2004 were evaluated by predicting water contents for the two following years 2005 and 2006. The simulated and measured water contents (Figure 11), in general, were in a good agreement for both years. An increase in water content at 5 and 20 cm depths after rainfall was predicted reasonably well. During a prolonged drought period, however, the simulated water contents decreased quicker than the measured values.

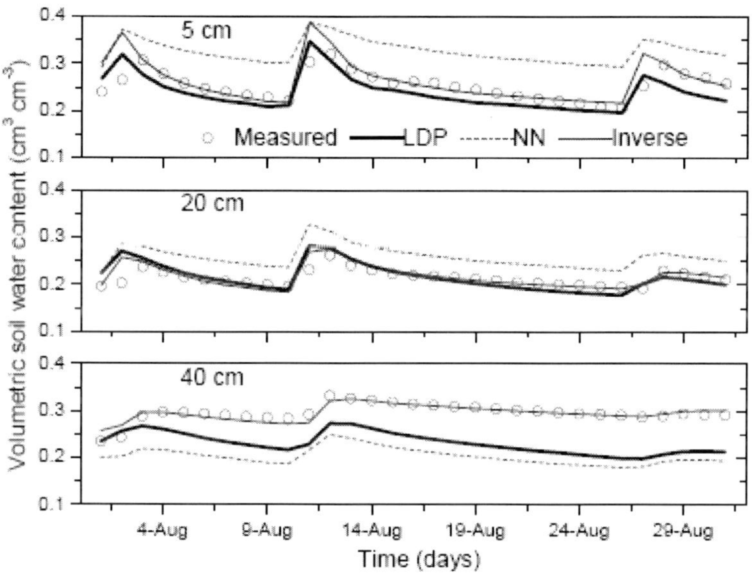

Figure 10. Comparison of the three simulation approaches with measured soil moisture in ungrazed since 1979 site in August 2004. LDP: laboratory-derived hydraulic parameters, NN: neural network, and Inverse: inverse optimization.

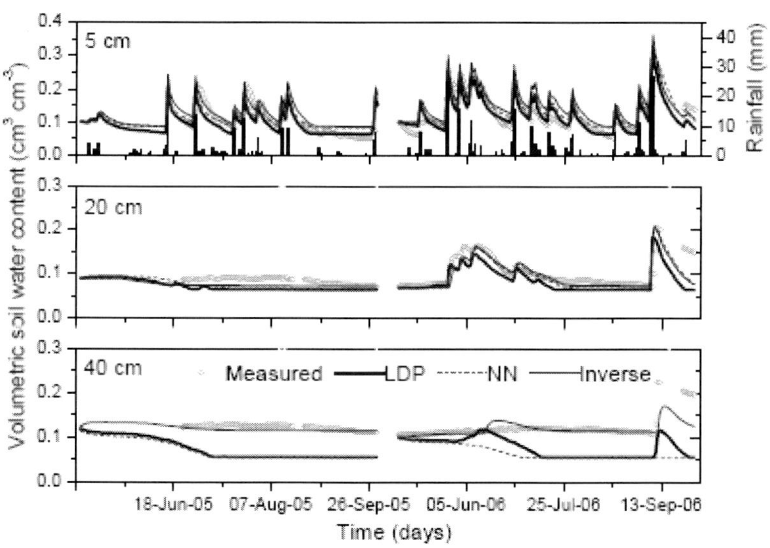

Figure 11. Soil moisture comparison between measured and simulated results in ungrazed since 1979 site during the growing periods in 2005-2006. LDP: laboratory-derived hydraulic parameters, NN: neural network analysis, and Inverse: inverse optimization.

Except for the inverse optimization, the simulated and measured water contents at 40 cm depth did not match well, which may be associated with an inaccurate field-based calculated *K(h)* in this layer. We found large differences in *K(h)* between the different model approaches at greater suction for the subsoil compared with the topsoil (Zhao *et al.,* 2010).

Table 6. Root mean square errors (*RMSE*) of three simulation approaches for model calibration and validation in three soil depths. LDP: laboratory-derived hydraulic parameter from soil cores, fitted by RETC software; NN: neural network values based on soil texture and bulk density analysis, predicted by ROSETTA; Inverse: simulation with the inverse optimization, initial parameter as LDP simulation

Site	Model Approach	August, 2004 (calibration)			May-September, 2005 (validation)			May-September, 2006 (validation)		
		5	20	40	5	20	40	5	20	40
		—— cm ——			—— cm ——			—— cm ——		
UG79	LDP	0.025	0.011	0.072	0.019	0.012	0.066	0.032	0.028	0.041
	NN	0.071	0.041	0.098	0.034	0.008	0.062	0.040	0.024	0.048
	Inverse	0.003	0.003	0.006	0.038	0.005	0.006	0.045	0.022	0.008
UG99	LDP	0.026	0.021	0.063	0.032	0.022	0.053	0.025	0.029	0.035
	NN	0.063	0.039	0.075	0.031	0.019	0.054	0.038	0.028	0.039
	Inverse	0.005	0.007	0.007	0.025	0.014	0.023	0.024	0.017	0.010
MG	LDP	0.041	0.020	0.051	0.026	0.051	0.016	0.042	0.044	0.035
	NN	0.043	0.031	0.067	0.033	0.059	0.021	0.052	0.045	0.039
	Inverse	0.011	0.012	0.040	0.025	0.045	0.019	0.033	0.034	0.032
HG	LDP	0.052	0.019	0.048	0.057	0.065	0.069	0.036	0.042	0.052
	NN	0.060	0.035	0.051	0.057	0.075	0.069	0.048	0.043	0.061
	Inverse	0.016	0.017	0.032	0.041	0.056	0.050	0.031	0.039	0.040

UG79: ungrazed since 1979, UG99: ungrazed since 1999, MG: moderately grazed only in winter time, and HG: heavily grazed.

The inverse method predicted the soil moisture profile best, as generally shown with the lowest RMSE (Table 6), because it effectively adjusted the laboratory measured parameters to fit the field data. The LDP did not yield satisfying predictions of the deeper soil layers. This approach did not account properly for the large pores measured in the laboratory and scale effects related to the sample size. The weakest prediction was obtained using the NN, which was likely due to the spatial variation of referred soil properties and the limited accuracy of PTFs themselves.

The NN model, compared with the LDP model, was especially incorrect at reflecting pore structural changes affected by e.g., soil compaction. As a consequence, predictions of water content were not adequate. This is in accordance with Børgesen and Schaap (2005), who indicated that to improve predictions of PTFs soil structural parameters should be included. Therefore, we considered that the in situ measured input parameters, when process-oriented modeling techniques were used, are essential to reflect the influence of land use change.

Model Prediction for Soil Water Fluxes

Based on these model calibration and validation steps, we used the inverse method to calculate water and energy budgets of the treatments. We expect that differences in evaporation at different sites should be attributed to differences in soil physical properties and soil surface coverage. The results showed that infiltrated water was mostly consumed by ET within a single growing period (Table 7). Grazing slightly decreased ET which could be explained by the sparse canopy structure and poorly structured soil induced by grazing. Under grazed conditions, this also led to greater net water storage and increased runoff. Compared with the two ungrazed sites, moderate grazing did not show clear effects on the water budget components, while heavy grazing decreased water interception from 17 to 7 mm and transpiration from 121 to 74 mm, and increased evaporation from 88 to 128 mm. The calculated mean Tp/ET ratios for the three growing seasons decreased gradually from 54% for UG79, to 50% for UG99, 48% for MG, and only 35% for HG. Therefore, while ET itself was slightly influenced by grazing, the components of ET significantly varied with grazing treatment. In the grazed sites, interception and transpiration decreased compared with the ungrazed sites where soil was covered more completely by live or dead plants. In contrast, evaporation in the grazed sites simultaneously increased due to larger bare soil surface. These findings are consistent with many other studies (e.g., Bremer *et al.*, 2001; Chen *et al.*, 2007).

Table 7. Simulated water budget components during growing periods in 2004-2006, inverse simulation result with Hydrus-1D (in mm, *P*: Precipitation, *I*: Interception, T_p: Transpiration, *E*: Evaporation, ΔS: Water storage change, *D*: Drainage, and *R*: Runoff; *Error* is model water balance errors)

Year	Treatment	*I*	T_p	*E*	ΔS	*D*	*R*	*Error*
2004	UG79	16.3	175.0	104.0	-24.6	1.0	1.7	1.6
(*P*= 275)	UG99	15.2	169.3	113.1	-26.3	3.2	1.6	-1.1
	MG	14.7	152.0	116.2	-18.0	3.1	4.1	2.9
	HG	7.1	125.2	142.3	-13.3	6.9	5.5	1.3
2005	UG79	14.8	71.3	59.4	1.3	2.4	0.1	-2.3
(*P*= 147)	UG99	14.4	65.4	63.4	5.5	1.3	0.2	-3.2
	MG	13.2	64.7	69.2	1.6	1.2	0.2	-3.1
	HG	6.6	40.3	90.6	7.4	5.1	0.5	-3.5
2006	UG79	18.3	117.3	100.3	7.3	1.2	0.8	-3.2
(*P*= 242)	UG99	17.4	107.7	112.3	4.3	1.0	0.5	-1.2
	MG	16.3	101.0	120.5	6.6	0.1	1.2	-3.7
	HG	8.2	56.8	151.7	16.8	2.0	4.6	1.9
Average	UG79	16.5	121.2	87.9	-5.3	1.5	0.9	-1.3
(*P*= 221)	UG99	15.7	114.1	96.3	-5.5	1.8	0.8	-1.8
	MG	14.7	105.9	102.0	-3.3	1.5	1.8	-1.3
	HG	7.3	74.1	128.2	3.6	4.7	3.5	-0.1

Average is the mean of three years. UG79: ungrazed since 1979, UG99: ungrazed since 1999, MG: moderately grazed only in winter time, and HG: heavily grazed.

Management Implication

The present study provided helpful insight into grazing management in a semi-arid steppe ecosystem. Our results proved that trampled ground leads to a more extreme environment than intact one. This was due to grazing impacts on soil and vegetation properties and hence water and energy partitioning. Whereas moderate to light grazing was in general less harmful, heavy grazing was detrimental to soil functions, which results in a reduced water storage (e.g., Greenwood and McKenzie, 2001; Martinez and Zinck, 2004). Meanwhile, live vegetation cover declined due to feeding and the mechanical effects of the hooves, and ET from the vegetated surface would be progressively replaced by soil evaporation. The insulating litter layer that would have restricted direct soil water loss would also fade off with increased grazing intensity (Zhao *et al.,* 2011a). Therefore, heavy grazing undoubtedly resulted in a drier soil water regime. Moreover, a shallower root distribution was observed in the HG site, which led to a reduction of the exploitable soil volume, a diminished amount of plant available water and thus a decreased rate of plant growth. Consequently, the heavy grazing site was definitely more susceptible to drought, and prone to soil degradation (Rietkerk and van de Koppel, 1997).

A reduction of grazing intensity could possibly increase soil water storage, and furthermore improve the system's productivity within a few years (Drewry, 2006). Our experiment suggests that the safe grazing intensity, like moderate grazing in this case, be about one to two sheep or sheep equivalents per ha for the grassland studied. However, projecting our short-term results to the long-term effects considering climate change and the feedback mechanisms (Li *et al.,* 2006), requires further research. This requests not only continuing flux measurements with a more intense field research, but also socioeconomic aspects of local land management. However, at this point we can strongly recommend reducing trampling intensity, and even animal exclusion in order to protect and restore degraded soils from intense grazing.

CONCLUSION

In this chapter we identified the main factors controlling the spatial patterns of soil moisture as a function of grazing intensity in a grazed semi-arid steppe ecosystem. Our results revealed that the variability of soil moisture was characterized by weak to moderate spatial structures depending on sampling time and grazing intensity. The correlation analysis showed that soil and plant properties, especially soil physical properties that were impacted by grazing intensity, were important in controlling the distribution of soil moisture. To adequately describe spatial patterns of soil moisture, we need to consider the dependency of controlling processes on the local conditions.

Using multivariate geostatistics, scale-dependency correlations between soil moisture and its controlling factors were detected. Soil and plant properties were proved to dominate the variation of soil moisture in UG99 at short-scale (45 m), in CG and HG at long-scale (90 m); but weakly controlled in UG79 and MG, suggesting that several controlling parameters might operate at the different spatial scales. This finding not only help to identify the mechanisms and processes acting on various scales, but provides the foundation for a coregionalization of physical and hydraulic parameters which subsequently may be used to improve ecohydrological models focusing on regional scales. Landscape discretization in ecohydrological units should take into account land management units and discontinuities, such as field distribution, soil and plant heterogeneity and the antecedent soil moisture.

Furthermore, based on soil pit scale long-term monitoring data, we divided a year into four hydric periods. In general, soil moisture in grazed sites was lower than in the ungrazed site, particularly for the 30-50 cm soil layer. Seasonal fluctuation of the soil moisture, due to variable precipitation and atmospheric demands, was most significant in the topsoil (0-10 cm) and was less pronounced in deeper soil. With increasing grazing intensity, soil water storage decreased remarkably. Consequently, grazing reduced plant available water and therefore grassland productivity, which are linked to a great extent with the trampling-induced soil structure change and soil moisture regime.

Finally, we showed grazing effects on eco-hydrological process and further parameterized HYDRUS-1D model to calculate coupled water and heat fluxes. Our conclusions are followings:

a) Grazing, especially heavy grazing, deteriorated soil functions, as indicated by decreased total and larger porosity accompanied by decreased saturated hydraulic conductivity and increased bulk density.
b) The hydraulic parameters presenting the soil structural functions simulate the soil water dynamics more accurately. As an example, laboratory-derived hydraulic parameter is better than parameter from neural network analysis method.
c) Water and temperature regimes were modified by grazing treatments, which was simulated best by the inverse optimization, followed by laboratory-derived hydraulic parameter and neural network analysis method.
d) The model results showed that grazing treatment increased soil heat fluxes. In comparison with the two ungrazed sites, moderate grazing did not show clear effects on the water budget, while heavy grazing significantly decreased interception and transpiration, and increased evaporation, runoff and drainage.
e) As a consequence, intense grazing not only reduced the amount of plant available water and thus grassland productivity, but also possibly increased the risks for wind and water erosion.

REFERENCES

Adler, P., and Lauenroth, W.K. (2000) Livestock exclusion increases the spatial heterogeneity of vegetation in Colorado shortgrass steppe. *Appl. Veg. Sci. 3*, 213–222.

Allen, R.G., Pereira L.S., Raes, D., and Smith, M., 1998. *Crop evapotranspiration–Guidelines for Computing Crop Water Requirements*. Irrigation and Drainage Paper No. 56, FAO, Rome, p 300.

Børgesen, C.D., and Schaap, M.G. (2005) Point and parameter pedotransfer functions for water retention predictions for Danish soils. *Geoderma* 127, 154–167.

Braun-Blanquet J., 1964. *Pflanzensoziologie*, Wien-New York: Springer Verlag.

Bremer, D.J., Auen, L.M., Ham, J.M., and Owensby, C.E. (2001) Evapotranspiration in a prairie ecosystem: Effects of grazing by cattle. *Agrono. J. 93*, 338–348.

Buttafuoco, G., Castrignano, A., Busoni, E., and Dimase, A.C., 2005. Studying the spatial structure evolution of soil water content using multivariate geostatistics. *J. Hydrol. 311*, 202–218.

Butterbach-Bahl, K., Kögel-Knabner, I., and Han. X.G., 2011. Steppe ecosystems and climate and land-use changes-vulnerability, feedbacks and possibilities for adaptation. *Plant Soil 340*, 1–6.

Cambardella, C.A., Moorman, T.B., Parkin, T.B., Karlen, D.L., Turco, R.F., and Konopka, A.E. (1994) Field scale variability of soil properties in Central Iowa soils. *Soil Sci. Soc. Am. J. 58*, 1501-1511.

Casa R., and Castrignanò A. (2007) Analysis of spatial relationships between soil and crop variables in a durum wheat field using a multivariate geostatistical approach. *European Journal of Agronomy*. doi:10.1016/j.eja.2007.10.001.

Casas, C., and Ninot, J.M. (2007) Soil water regime through contrasting pasture communities in a Sub-mediterranean landscape. *J. Hydrol. 335*, 98–108.

Chen, Y., Lee, G., Lee, P., and Oikawa, T. (2007) Model analysis of grazing effect on above-ground biomass and above-ground net primary production of a Mongolian grassland ecosystem. *J. Hydrol.* 333(1), 155–164.

Chen, Z.Z., Wang, S.P. (2000) *Chinese Typical Grassland Ecosystem*. China Science Press, pp. 106–109.

Chung, S., and Horton, R. (1987) Soil heat and water flow with a partial surface mulch. *Water Resour. Res.* 23, 2175–2186.

Deutsch, C.V., and Journel, A.G. (1998) *GSLIB Geostatistical Software Library and Users Guide*. Oxford University Press: Oxford.

Drewry, J.J. (2006) Natural recovery of soil physical properties from treading damage of pastoral soils in New Zealand and Australia: a review. *Agr. Ecosyst. Environ. 114*, 159–169.

Feddes, R.A., Kowalik, P.J., and Zaradny, H. (1978) Simulation of field water use and crop yield, John Wiley and Sons, New York, NY.

Flerchinger, G.N., Sauer, T.J., and Aiken, R.A. (2003) Effects of crop residue cover and architecture on heat and water transfer. *Geoderma 116*, 217–233.

Francis C.F., Thornes, J.B., Romero Diaz, A., Lopez Bermudez, F., and Fisher, G.C. (1986) Topographic control of soil moisture, vegetation cover and land degradation in a moisture stressed Mediterranean environment. *Catena 13*, 211–225.

Gao, Y.Z., Giese, M., Lin, S., Sattelmacher, B., Zhao, Y., and Brueck, H. (2008) BeloMGround net primary productivity and biomass allocation of a grassland in Inner Mongolia is affected by grazing intensity. *Plant Soil 307*, 41–50.

Gómez-Plaza, A., Martínez-Mena, M., Albaladejo, J., and Castillo, V.M. (2001) Factors regulating spatial distribution of soil water content in small semiarid catchments. *J. Hydrol. 253*, 211–226.

Goovaerts, P. (1992) Factorial kriging analysis–a useful tool for exploring the structure of multivariate spatial soil information. *J. Soil Sci. 143*, 597–619.

Goovaerts, P. (1998) Geostatistical tools for characterizing the spatial variability of microbiological and physico-chemical soil properties. *Biol. Fertil. Soils 27*, 315-334.

Grayson, R.B, Blöschl, G., Western, A.W., and McMahon, T.A. (2002) Advances in the use of observed spatial patterns of catchment hydrological response. *Adv. Water Resour. 25*, 1313–1334.

Grayson, R.B., Western, A.W., and Chiew, F.H.S. (1997) Preferred states in spatial soil moisture patterns: Local and nonlocal controls. *Water Resour. Res. 33*, 2897–2908.

Green, T.R., Ahuja, L.R., and Benjamin, J.G. (2003) Advances and challenges in predicting agricultural management effects on soil hydraulic properties. *Geoderma 116*, 3–27.

Greenwood, K.L., and McKenzie, B.M. (2001) Grazing effects on soil physical properties and the consequences for pastures: a review. *Austr. J. Exp. Agr. 41*, 1231–1250.

Hansson, K., Šimůnek, J., Mizoguchi, M., Lundin, L.-C., and van Genuchten, M.T. (2004) Water flow and heat transport in frozen soil: numerical solution and freeze-thaw applications. *Vadose Zone J. 3*, 693–704.

Hawley, M.E., Jackson, T.J., and McCuen, R.H. (1983) Surface soil moisture variation on small agricultural watersheds. *J. Hydrol. 62*, 179–200.

Hébrard, O., Voltz, M., Andrieux, P., and Moussa, R. (2006) Spatio-temporal distribution of soil surface moisture in a heterogeneously farmed Mediterranean catchment. *J. Hydrol. 329*, 110–121.

Johnston, K., Ver Hoef, J.M., Krivoruchko, K., and Lucas, N. (2001) *Using ArcGIS Geostatistical Analyst*. ESRI, New York.

Kölbl, A., Steffens, M., Wiesmeier, M., Hoffmann, C., Funk, R., Krümmelbein, J., Reszkowska, A., Zhao, Y., Peth, S., Horn, R., Giese, M., and Kögel-Knabner, I. (2011) Grazing changes topography-controlled topsoil properties and their interaction on different spatial scales in a semi-arid grassland of Inner Mongolia, P.R. China. *Plant Soil 340*, 35–58.

Krümmelbein, J., Wang, Z., Zhao, Y., Peth, S., and Horn, R. (2006) Influence of various grazing intensities on soil stability, soil structure and water balance of grassland soils in Inner Mongolia, P. R. China. Pp: 93-101. In: *Soil Management for Sustainability*. Horn, R, Heiner, F., Peth, S. Peng, X. H., (eds.), Catena Verlag GMBH, 33447 Reiskirschen, Germany.

Kutilek, M., and Nielsen, D.R. (1994) *Soil Hydrology. Catena Verlag*, 560 Cremlingen, pp. 70–84.

Leenhardt, D., Voltz, M., and Rambal, S. (1995) A survey of several agroclimatic soil water balance models with reference to their spatial application. *Eur. J. Agron. 4*,1–14.

Li, S.G., Eugster, W., Asanuma, J., Kotani, A., Davaa, G., Oyunbaatar, D., and Sugita, M. (2006) Energy partitioning and its biophysical controls above a grazing steppe in central Mongolia. *Agr. Forest Meteorol. 137*, 89–106.

Li, S.G., Harazono, Y., Oikawa, T., Zhao, H.L., He, Z.Y., and Chang, X.L. (2000) Grassland desertification by grazing and the resulting micrometeorological changes in Inner Mongolia. *Agric. For. Meteorol. 102*, 125-137.

Lin, H. (2006) Temporal stability of soil moisture spatial pattern and subsurface preferential flow pathways in the Shale Hills catchment, *Vadose Zone J. 5*, 317–340.

Loos, C., Gayler, S., and Priesack, S. (2007) Assessment of water balance simulations for large scale weighing lysimeters. *J. Hydrol. 335*, 259–270.

Martinez, L.J., and Zinck, J.A. (2004) Temporal variation of soil compaction and deterioration of soil quality in pasture areas of Colombian Amazonia. *Soil Till. Res. 75*, 3–17.

Ndiaye, B., Molénat, J., Hallaire, V., Gascuel, C., and Hamon, Y. (2007) Effects of agricultural practices on hydraulic properties and water movement in soils in Brittany (France). *Soil Till. Res. 93*, 251–263.

Reszkowska, A., Krümmelbein, J., Peth, S., Horn, R., Zhao, Y., and Gan, L. (2011) Influence of grazing on hydraulic and mechanical properties of semiarid steppe soils under different vegetation type in Inner Mongolia, China. *Plant Soil 340*, 59–72.

Rietkerk, M., and van de Koppel, J. (1997) Alternate stable states and threshold effects in Semi-Arid grazing systems, *Oikos 79*, 69–76.

Ritchie, J.T. (1972) Model for predicting evaporation from a row crop with incomplete cover. *Water Resour. Res. 8*, 1204–1213.

Schaap, M.G., Leij, F.J., and van Genuchten, M.T. (2001) Rosetta: A computer program for estimating soil hydraulic parameters with hierarchical pedotransfer functions. *J. Hydrol. 251*, 163–176.

Sneath, D. (1998) State policy and pasture degradation in Inner Asia. *Science 28*, 1147–1148.

Snyman, H.A. (2005) Rangeland degradation in a semi-arid South Africa - I: influence on seasonal root distribution, root/shoot ratios and water-use efficiency. *J. Arid Environ. 60*, 457–481.

Šimůnek, J., Sejna, M., and van Genuchten, M.Th. (1998) *The HYDRUS-1D software package for simulating the one dimensional movement of water, heat, and multiple solutes in variably-saturated media.* Version 2.0. IGWMC-TPS-70. Int. GroundWater Modeling Center, Colorado School of Mines, Golden.

Steffens, M., Köbl, A., Totsche, K.U., and Kögel-Knabner, I. (2008) Grazing effects on soil chemical and physical properties in a semiarid steppe of Inner Mongolia P.R. China. *Geoderma 143*, 63–72.

Sugita, M., Asanuma, J., Tsujimura, M., Mariko, S., Lu, M.-J., Kimura, F., Azzaya, D., and Adyasuren, T. (2007) An overview of the rangelands atmosphere–hydrosphere–biosphere interaction study experiment in northeastern Asia (RAISE). *J. Hydrol. 333*, 3–20.

van Genuchten, M.Th. (1980) A closed-form equation for predicting the hydraulic conductivity of unsaturated soils. *Soil Sci. Soc. Am. J. 44*, 892–898.

van Genuchten, M.Th., Leij, F.J., and Yates, S.R. (1991) *The RETC code for quantifying the hydraulic functions of unsaturated soils.* Ver. 1.0 EPA Rep. 600/2-91/065. U.S. Salinity Lab., USDA-ARS, Riverside, CA.

Willat, S.T., and Pullar, D.M. (1983) Changes in soil properties under grazed pastures. *Austr. J. Soil Res. 22*, 343–348.

Western, A.W., Grayson, R.B., Blöschl, G., Willgoose, G.R., and McMahon, T.A. (1999) Observed spatial organisation of soil moisture and its relation to terrain indices. *Water Resour. Res. 353*, 797–810.

World Resources Institute (2000) *World Resources 2000–2001. People and Ecosystems: The Fraying Web of Life.* World Resources Institute, Washington, DC, USA.

Zhao, Y., Peth, S., Horn, R., Krümmelbein, J., Ketzer, B., Gao, Y.Z., Doerner, J., Bernhofer, C., and Peng, X.H. (2010) Modeling grazing effects on coupled water and heat fluxes in Inner Mongolia grassland. *Soil Till. Res. 109*, 75–86.

Zhao, Y., Peth, S., Horn, R., Hallett, P., Wang, X.Y., Giese, M., and Gao, Y.Z. (2011a) Factors controlling the spatial patterns of soil moisture investigated by multivariate and geostatistical analysis. *Ecohydrology 4*, 36–48.

Zhao, Y., Peth, S., Reszkowska, A., Gan, L., Krümmelbein, J., Peng, X., and Horn, R. (2011b) Response of soil moisture and temperature to grazing intensity in a Leymus chinensis steppe, Inner Mongolia. *Plant Soil 340*, 89–102.

In: Grazing Ecology: Vegetation and Soil Impact
Editor: José Ramón Arévalo
ISBN 978-1-62100-436-3

Chapter 8

DIVERSITY IN MANAGEMENT REGIMES: A MAJOR DRIVER OF THE PRODUCTION / CONSERVATION TRADE-OFFS IN GRASSLAND AGROECOSYSTEM

***Rodolphe Sabatier*[*1,2], *Luc Doyen*[3], *Francois Leger*[1,2] and *Muriel Tichit*[1,2]**
[1]INRA, UMR 1048 SAD APT, F-75005 Paris, France
[2]AgroParisTech, UMR 1048 SAD APT, F-75005 Paris, France
[3]CNRS, UMR 7204 CERSP, MNHN, F-75005 Paris, France

ABSTRACT

Over the last two decades, agri-environment schemes have been developed by member states of the European Union. Recent assessments reveal the lack of success of these schemes in stopping biodiversity loss. While initial schemes focused on wildlife-friendly practices at field scale, the agro-landscape scale is increasingly considered as the relevant scale of analysis and intervention. Recent studies argue that efficient schemes require modifying both management practices at field scale and their spatial arrangements at landscape scale, in order to increase the spatial heterogeneity of agro-landscapes. To date however, only a few studies quantified the relationships between agricultural production and biodiversity conservation in farmlands. The objective of this chapter is to asses, at different spatial and temporal scales the trade-offs between production and bird conservation in a grassland agroecosystem used for meat cattle breeding. The research methodology is based on several models, built within the mathematical framework of the viability theory. These models formalize the interactions between grazing and mowing practices and the ecological dynamics of two grassland bird species at three nested spatial scales (field, farm and landscape). The five major findings of this thesis are as follows: at field scale (1) The timing and intensity of grazing are the major drivers of the trade-off between production and bird conservation, (2) the best ecological performances are obtained for intermediate levels of productive performances, (3) the inter-annual variations of management practices, made possible by result-oriented agri-environment schemes, improve ecological performances while at the same time

* Email: rsabati@gwdg.de

increase grassland management flexibility, (4) at farm scale, the proportion of management regimes is a major driver of the trade-off between production and conservation, but bird conservation always implies a production cost, (5) at landscape scale, the spatial arrangement of different land uses improves the trade-off between production and conservation. Overall, these results reveal the different drivers of the production / conservation trade-offs at different scales. At each scale, the interactions between the different management regimes are the basis of the trade-off mechanisms between production and conservation. The importance of the spatial arrangement of management regimes suggests that the conciliation between production and conservation could strongly benefit from increased coordination among farms.

Keywords: Agri-environments, conservation, grasslands, grazing models

INTRODUCTION

Beyond the social demand for biodiversity conservation, new paradigms in agricultural sciences tend to put biodiversity back in the centre of productive dynamics (Altieri and Nicholls, 2005). Biodiversity conservation becomes therefore more than a constraint but a necessity for the set up of sustainable systems. Grassland management by farmer has complex effects on biodiversity. In the one hand, as in most other agricultural areas, grassland biodiversity strongly suffers from agricultural intensification (Vickery *et al.,* 2001). In the other hand, these ecosystems strongly depend on farming activities that maintain them open (e.g. Gibon *et al.,* 2010). Biodiversity conservation in grassland areas therefore appears to be a major challenge of multifunctional management of grasslands (Brussard *et al.,* 2010). Such a multi functionality leads to trade-offs between agricultural production and biodiversity conservation. Knowing the shape of these trade-offs brings highly relevant information to define conservation policies (Green *et al.,* 2005).

The aim of this chapter is to give an evaluation of the multi-scale trade-offs that occur between productive and ecological performances in a grassland agroecosystem. More precisely, we raised two questions:

1) What is the shape of the relationship between productive and ecological performances?
2) Is it possible to overcome the trade-off that arises between production and conservation?

The grassland agroecosystem is a multi-scale system (Figure 1). These two questions had therefore to be considered at field scale and at upper sales. Field scale is coherent on both ecological and productive dimensions. It is the scale at which the first stages of the bird life cycle occur. On the productive point of view, this scale is the elementary unit. At this scale, we considered management as homogeneous. Both ecological and productive dynamics also depend on upper scales that correspond to an aggregation of different fields. Considering these larger sets of fields, there is no single scale for both ecological and productive dimensions. On the one hand, farm scale is coherent on the productive point of view but do not necessarily correspond to a continuous space. On the other hand, landscape is a continuous area, coherent on the ecological point of view but is generally composed of several farms. We therefore considered both scales. First, we defined the shape of the trade-off at farm scale, then we looked for a way to overcome the trade-off at landscape scale where we could integrate more complex ecological dynamics.

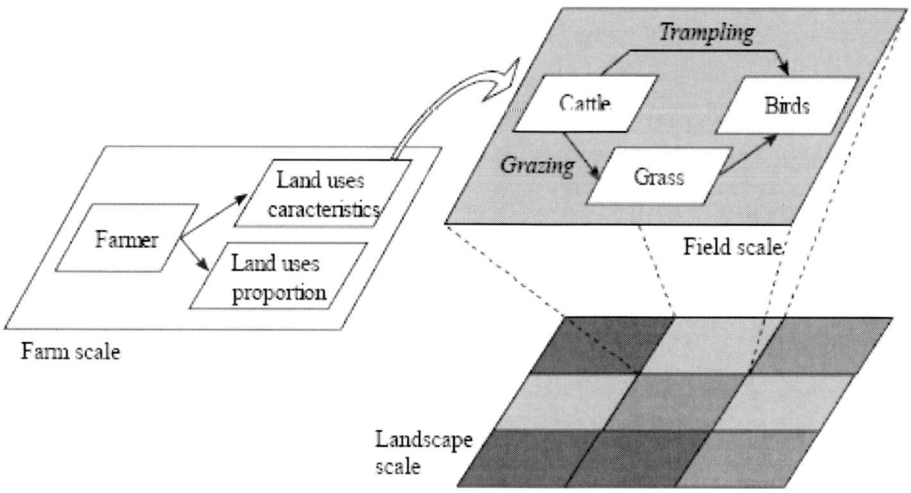

Figure 1. Overview of the grassland agroecosystem with a special focus on bird conservation in grazed grasslands. Three scales are mentioned: farm field and landscape scale. Each colour on the grey scale corresponds to a different land use.

As a case study, we focused on the conservation of Lapwings *Vanellus vanellus* in wet grasslands of the French Atlantic coast (46°22’ N, 1°25’ W). Wet grasslands are strongly shaped by farming practices such as grazing or mowing. These agroecosystems are therefore challenging systems to study the effect of agricultural management on biodiversity. Due to their high position in trophic networks and their close connection with wet grasslands, wader species give good information about the health of the ecosystem (Flint, 1998). The Lapwing life cycle is deeply linked to the management of grassland (Milsom *et al.,* 2000; Durant *et al.,* 2008b). At field scale they are impacted by management in both direct and indirect ways. Nest trampling by cattle destroys eggs and has a direct effect on bird fecundity (Beintema and Muskens, 1987). Grazing modifies sward height and sward structure that are key parameters of habitat quality which has an indirect effect on juvenile survival (Durant *et al.,* 2008a, b). Lapwing dynamics are also impacted by upper scale dynamics. Juveniles are nidifugous, which means that they move from their site of birth to a breeding site a few hundred meters away (Redfern, 1982). The proportion of the different land uses (and of the habitats that they generate) as well as the structure of the landscape may impact their dynamics.

Such a multi-scale study of an agroecosystem was made possible using a dynamic modelling approach. In this chapter we will present a synthesis of different studies made by our team. After briefly presenting the main aspects of the models, we will present the main results and discuss them with a specific focus on the role played by management diversity on the trade-offs between agricultural production and biodiversity conservation.

MULTI-SCALE MODELLING

Model Overview

We built three dynamic models at field, farm, and landscape scales. Each model relies on a state-control approach (Clark and Mangel, 2000) that represents a grassland agroecosystem which is the breeding habitat of a bird species, the Lapwing, and the feeding resource for domestic cattle. The models are discrete time matrix models (Caswell, 1989) linking grass dynamics to bird population dynamics. Time step is defined on a monthly basis as most farmers in the study area implement middle term grazing sequences (three weeks to several months). In the grazed grass sub-model, biomass is harvested through grazing or mowing. The bird sub-model simulates population changes over time in response to the direct and indirect effects of grazing and mowing on bird life traits. At field scale, grazing intensity has a direct effect on clutch size due to nest trampling by cattle. Mowing occurs after bird birth and has no effect on their fecundity. Sward height (i.e. habitat quality) strongly depends on grazing and mowing and impacts juvenile survival. At upper scales, bird movements depending on habitat qualities impact their dynamics.

The models may compute indicators summarizing the ecological and productive performance of each management strategy as well as the overall flexibility of the system.

Numerical computations were performed with the Scilab 4.1.3 software (http://www.scilab.org/ ; Scilab Consortium 2007).

DYNAMICS

Field Scale Dynamics

Sward dynamics are modelled using matrix based dynamics. The grazed sward dynamics represent the temporal dynamics of biomass controlled by grazing and mowing. It includes sward growth, senescence and decay rates that are time dependent on a monthly basis. The state of the biomass is linked to sward height through a linear function. For further details on biomass dynamics and parameter values, see Sabatier *et al.* (2010).

The second state of the system describes the Lapwing population. Once again, dynamics are formalised by a matrix model in discrete time on a monthly basis. These dynamics include direct and indirect effects of management on bird dynamics. Bird fecundity is linked to grazing intensity through a decreasing function (high grazing intensity implies low bird fecundity). Bird survival is linked to sward height through a threshold function (maximal survival within the range of suitable sward heights and low survival out of this range). For further details concerning bird dynamics see Sabatier *et al.* (2010).

Upper Scales Dynamics

In addition to these field scale dynamics, upper scale processes are integrated in the farm and landscape scale models. In this section, we will give a qualitative description of the processes at stake. For all the quantitative details concerning these processes see Sabatier (2010).

Viability analyses

The first two models (field and farm scales) were built under the mathematical framework of the viability theory (De Lara & Doyen, 2008) which is closely related to the viability theory (Aubin 1991). This framework enables the satisfaction of production, socio-economic and environmental constraints and is, in this respect, a multi-criteria approach that is very useful in environmental studies. It makes it possible to find the whole set of viable management strategies that keep a system within some constraints.

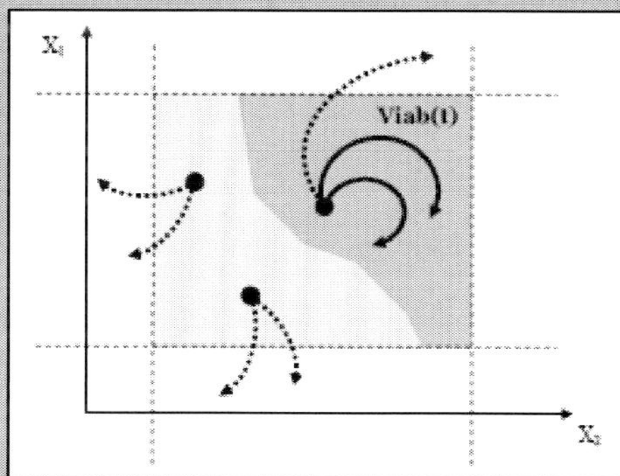

Concepts linked to the viability theory in the case of a two state (X_1X_2) system. Dotted lines stand for the different constraints of the system. Arrows stand for the different trajectories. The ones that respect constraints through time (continuous) are said viable, the other (dotted) are unviable. Viability algorithm make it possible to look for the so called viability kernel (Viab(t)) which is the set of states from which there is at least one viable trajectory.

Different sets of constraints were defined for each model corresponding to ecological and productive aims as well as biotechnical limits of feasibility.

Ecological constraints

At field scale, ecological constraints were a cattle density threshold limiting trampling during the nesting month and a sward height threshold ensuring a good quality habitat during the rearing month. These constraints were replaced by a bird population threshold when considering result-oriented measures. An additional ecological constraint was defined at farm scale, ensuring a minimal proportion of so called ecological grazing within the farming system. This ecological grazing is a grazing sequence that respects ecological constraints at field scale.

Productive constraints

A productive constraint defines a minimal level of productive performance necessary for the farmer. The productive performance corresponds to the number of grazing days (simplifying to 30 days per month) associated with a grazing strategy. The productive constraint was only applied when comparing habitat and result-oriented scenarios. The value of the threshold was defined by the 10% lower quantile of a dataset of 344 real grazing strategies recorded on our study site (Durant *et al.*, 2008a).

Constraints linked to the biotechnical feasibility

The first viability constraint linked to the biotechnical feasibility concerns all grazing regimes. It is related to the satisfaction of cattle feeding requirements and restricts cattle density so that biomass needs do not excess the available biomass. At farm scale a second constraint is defined, it ensures that the total number of animals effectively grazing at time t do not exceed the size of the herd.

The farm scale model includes the spatial component as proportions of land uses. In the landscape model, in addition to this proportion, the spatial position of land uses in the landscape is also included.

At both farm and landscape scales, three types of land uses are defined: ecological grazing (EG), productive grazing (PG), and mowing (M). These three land uses correspond to different functions of the grassland. Ecological grazing corresponds to a grazing sequence suitable for birds whereas productive grazing only maximizes harvest by cattle. Mown grasslands provide hay. On the productivity aspect, these three land uses interact within the farming system. In the farm model, hay is used to feed cattle when it is not feeding on grasslands. In the landscape model, no explicit productive link exists between the different land uses as the different fields of the landscape may not belong to the same farm.

On the ecological point of view, both models include bird movement. Movements are conditioned by habitat quality (sward height) generated by the different land uses. In the farm model, movements only depend on the proportion of land uses and concern juveniles and adults. In the landscape model, focus is given on juvenile movements that also depend on spatial structure of the landscape, birds being unable to move too far from their site of birth.

SIMULATIONS

At field scale a first step was to draw the shape of the trade-off between production and conservation. To do so, we assessed the ecological and productive performances of different grazing strategies corresponding to contrasting sets of viability constraints while maximizing or minimizing the productive performance. Productive performance was the number of grazing days in livestock units (LU)$\cdot$ha$^{-1}\cdot$year^{-1} and ecological performance was the bird population size at time horizon.

Then we looked for a way to overcome the trade-off between production and conservation using result-oriented agri-environmental schemes. We compared the performance of the system for habitat and result oriented scenarios. In the habitat-oriented scenario, the ecological constraints correspond to trampling and sward height thresholds whereas in the result-oriented scenario, only a bird population threshold is used. Ecological and productive indicators were similar to that used to draw the trade-off at field scale. The flexibility was assessed using the size of the viability tubes which is a good indicator of the number of viable strategies. For further details on the viability tubes see Sabatier (2010).

At farm scale, the trade-off curve was drawn plotting the ecological and productive performances of management strategies maximising production for various values of the minimal proportion of ecological grazing. The ecological performance was the bird population size and the productive performance was the total biomass harvested.

At landscape scale, we generated a wide set (n=22,356) of landscapes that had a constant proportion of mowing, identical to that observed on the study site (39%; Sabatier *et al.* 2008) and various proportions of ecological and productive grazing. Each landscape of this sub-set was characterized by its ecological and productive performances, by the proportion of the different land uses, and by a landscape structure index. The productive performance of agro-landscapes is obtained with a grazed grassland production index (*P*) that corresponds to the number of days for which feed for cattle is provided by grasslands, expressed in LU$\cdot$days$\cdot$ha^{-1} of grazed pastures. The productive performance index had the property of being a linear function of the PG/EG ratio. Ecological performance is the bird population size after one year.

RESULTS

At Field Scale, Timing and Intensity of Grazing Are Major Drivers of the Trade-Off

The trade-off showed a non linear shape and best ecological performances were reached at intermediate levels of productive performances (Figure 2).

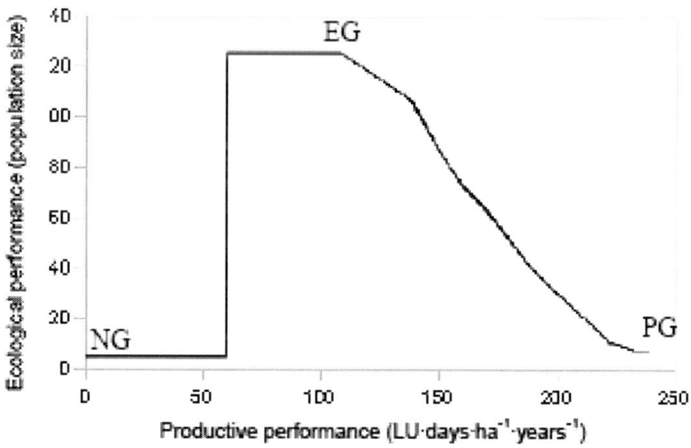

Figure 2. Trade-off between agricultural production and biodiversity conservation at field scale with different extreme scenarios (NG: No Grazing, EG: Ecological Grazing, PG: Productive grazing).

The trade-off at field scale shows a non monotonous shape. With the no-grazing (NG) scenario ecological performance was low since suitable sward heights could not be obtained and population sizes at 15 years were small (Figure 2).

The viable scenario based on minimal grazing revealed that a minimum of 60 grazing days per year was compulsory in order to maintain the populations over the long term (Figure 2). This scenario generated viable sward heights ensuring high chick survival. It was characterised by a late summer grazing sequence without any spring grazing and thus no trampling. Heavy summer grazing had delayed effects on grass biomass in the following early spring where biomass was 70% lower than in the no-grazing scenario. Such low levels of biomass constrained biomass growth and made it possible to maintain sward heights within the sward height corridor.

In the viable scenario based on maximal grazing (EG), productive performance reached 108 grazing days per year and ecological performance led to high population sizes at 15 years (Figure 2). Besides autumn grazing, the grazing strategy also implied a grazing peak in June, after the critical month for nest trampling (Figure 3). Suitable sward heights for birds were mainly generated thanks to autumn grazing. The late spring grazing sequence took advantage of spring growth from the productive viewpoint whereas in the previous scenario, the available biomass was not grazed and thus lost through decay.

Above 108 grazing days, no grazing strategy could maintain the population to the higher level. Between 108 and 240 grazing days, the ecological-productive relationship showed a Pareto like frontier. Any improvement in productive performance entailed a decrease in population size. Symmetrically an increase in population size implied a decrease in the annual number of grazing days.

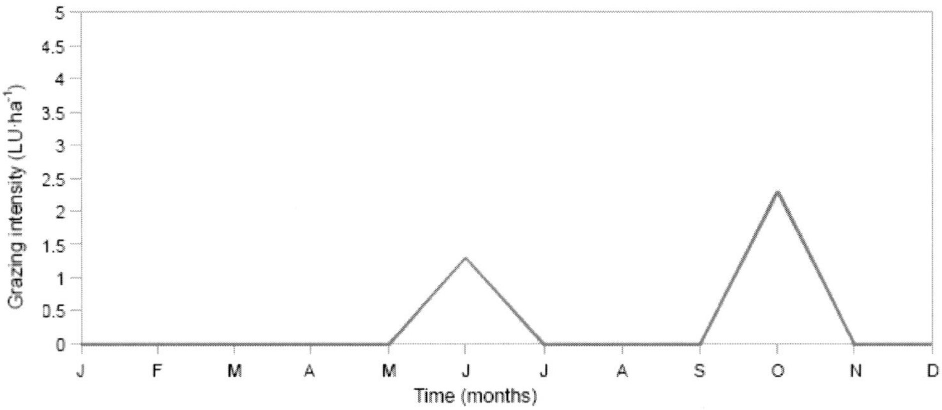

Figure 3. Grazing sequence of the ecological grazing.

As expected, the productive scenario (PG) led to the highest productive performance (≈ 240 grazing days) (Figure 4). It was reached with grazing strategies leading to high trampling in April, therefore, bird populations were heavily impacted (Figure 2).

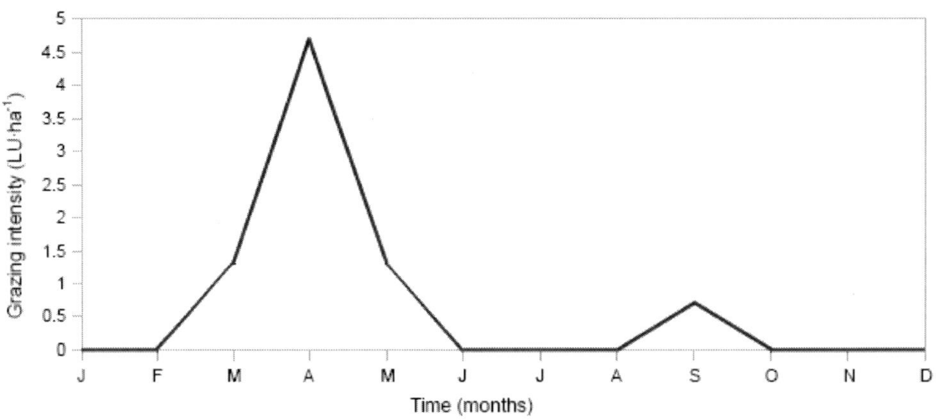

Figure 4. Grazing sequence of the productive scenario PG.

Result-Oriented Schemes Mainly Improve the Flexibility of the System

Figure 5.a shows the average ecological and productive performance of a sample of 10 000 grazing strategies for each of the two habitat and result oriented scenarios for initial bird population size N(0)=30. Comparison of both average ecological and productive performance of the two scenarios showed significant differences (T-test, P-value<0.001). However, differences were low, especially regarding productive performances.

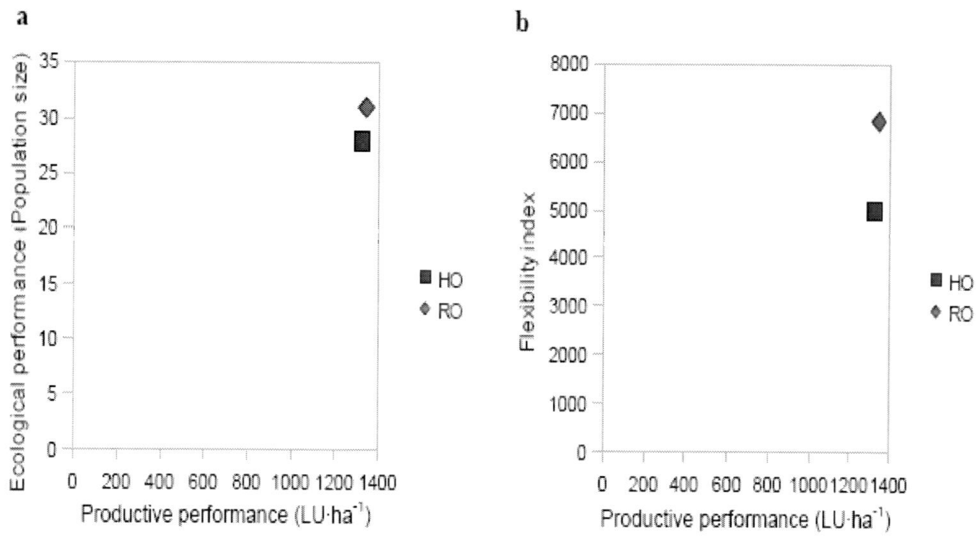

Figure 5. Comparison of the performances obtained with the habitat-oriented (HO) and the result-oriented (RO) scenarios. Figure a shows the ecological and productive performances, Figure b shows the productive performance and the flexibility index.

Figure 5.b shows the average productive performance compared to an indicator of its flexibility. Figure 5.b shows that the difference in flexibility between the two scenarios is much higher than the difference in productive performance. Result-oriented scenarios allowed the farmer to implement much more viable strategies and led to a much higher flexibility of the system.

At Upper Scales the Proportion of Land Uses is a Strong Driver of the Trade-Off

For various proportions of ecological grazing, the model revealed the whole trade-off between productive and ecological performances at farm scale (Figure 6). It was not possible to maximize both ecological and productive performances. A trade-off corresponding to a set of Pareto optimum states emerged, where any ecological improvement involved a production loss and vice versa. However, the trade-off was convex meaning that productive and ecological performances were always higher than the weighted sum of the management regime performances.

This trade-off was driven by the proportion of different land uses. When maximizing grassland production (Figure 6, farm A), no area was allocated to ecological grazing. The farm area was roughly equally allocated to mowing and productive grazing. Ecological performance was nearly nil with such "productive proportions". In order to maintain bird populations, a given proportion of the farm area had to be allocated to ecological grazing (Figure 6, farm B).

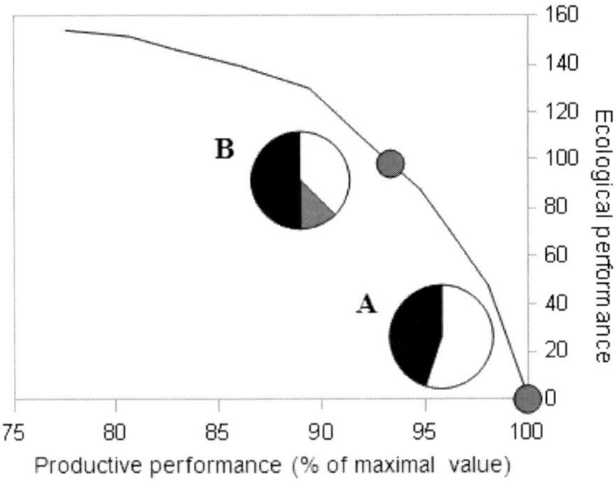

Figure 6. Trade-off between ecological and productive performances at farm scale. Proportion of land uses are given for two particular points of the trade-off: farm aimed at production only (A) and farm aimed at maintaining bird populations to 100 (B). Productive grazing is in black, ecological grazing is in grey, and mowing is in white.

Landscape Structure Makes It Possible to Overcome the Relationship Trade-Off

At the landscape scale, a strong relationship existed between productive performance and ecological performance (Figure 7; Anova $p<0.1$, $R^2=0.20$). The linear relationship between the grazed land uses proportion and the productive performance means that there is also a strong relationship between ecological performance and the proportion of the two types of grassland (PG and EG).

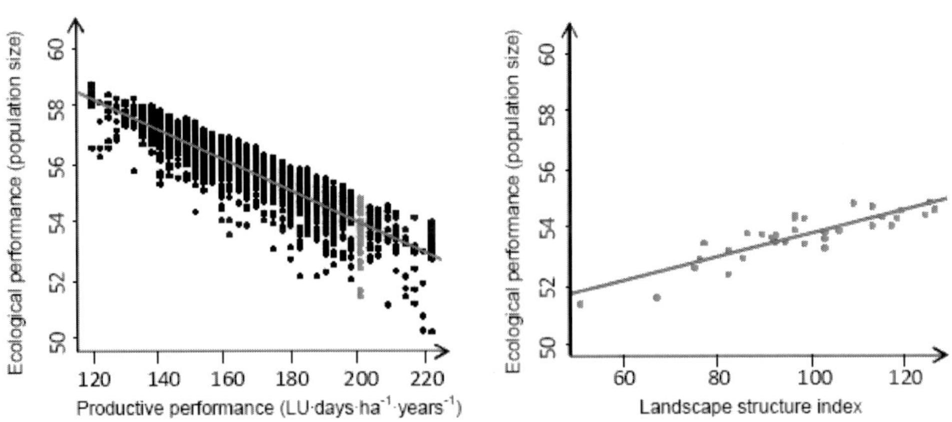

Figure 7. Effect of land use proportion and landscape structure on the ecological and productive performance of grassland landscapes with three land uses. Mowing proportion is set to 38 to 40%. The landscapes in grey correspond to landscapes composed of 9 to 11% of ecological grazing EG. Regression curves are in grey.

There was nevertheless a large part of variance that was not explained by the proportion of land use. For a constant proportion of ecological grazing (e.g. 10%; +/- 1% in the example in Figure 7), this residual variance was strongly explained by the landscape structure (Anova $p<0.001$, $R^2=0.75$). For a constant proportion of land uses, an increase in the complexity of spatial arrangement of land uses in the landscape was a way of improving its ecological performance without losing on the productive dimension. The spatial structure of the landscape therefore proved to be effectively a good way to overcome the relationship between production and conservation.

DISCUSSION

Diversity: Central Mechanism in the Reconciliation of Production and Conservation

The Production-Conservation Trade-Off Depends on a Choice of Management Regime

Grassland landscapes are composed of sets of fields on which farmers operate different management regimes. Here, we give an ecological point of view on the concept of management regime developed by the community of livestock science. A management regime can be defined as a rational succession of harvesting periods through the year, each characterized by a harvesting method (grazing or mowing) and a given intensity (Beranger and Micol, 1981). The choice for the farmer to assign a management regime to a given field depends on both its feeding strategy and on "tactical" decisions related to the adequacy of this feeding strategy to environmental conditions (mainly climatic ones) (Guerin *et al.*, 1994). This section refers particularly to two specific management regimes (Figure 8).

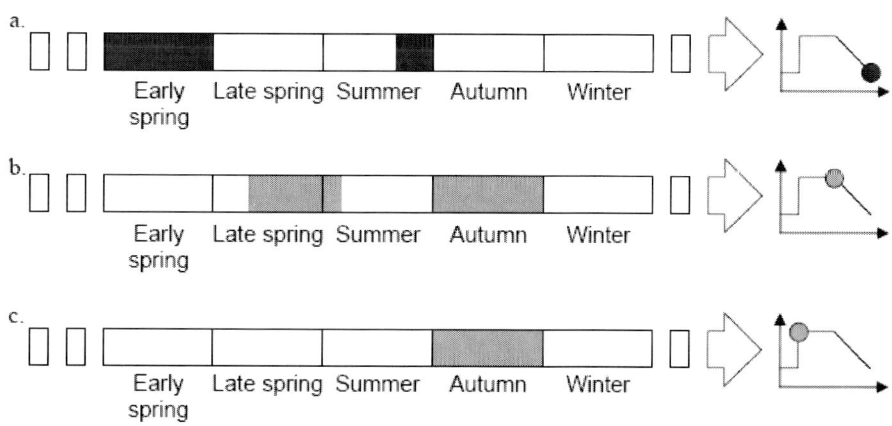

Figure 8. Productive (a, dark grey) and ecological (b, c, light grey) management regimes. Grazing periods are in colour, periods without grazing are white. Management regimes b and c correspond to the two possible ecological grazing. Each management regime is linked to its position on the relationship between productivity performance and ecological performance as defined on figure 2.

The productive management regime is the one that maximizes productive performances without ecological constraint. It maximizes harvest through an early spring grazing linked to a late summer grazing of regrown biomass. On the ecological point of view, it implies high trampling and therefore low ecological performance.

The ecological management regime b (and c) maximizes productive under ecological constraints. It is a management regime that combines an autumn grazing to an optional late spring grazing. These two grazing periods have distinct purposes:

- Although the amount of biomass available for grazing is lower in autumn than in spring, autumn grazing provides a cheap food source in times of low food availability. The value of the forage is not the same in spring where the biomass is an abundant resource on all types of grassland, and in autumn, where it is rarer but plays an important role in the production system. It makes it possible to delay wintering of cattle and thus to delay the need of dry forage.
- Autumn grazing leads to a strong reduction of sward height, which impacts sward growth the next spring. This delayed effect results in delaying the biomass production peak in late spring. For the farmer, this means a later exit of animals but also greater food availability in late spring. The variations of this management regime only differ by the presence or absence of late spring grazing. In ecological terms this management regime is favourable because it makes it possible to generate suitable sward height while limiting spring trampling.

It is therefore the management regimes as a whole that are the basis of the trade-off between production and conservation. In other words, this trade-off does not depend on a particular isolated grazing period but rather on the interaction between diversified grazing periods within a management regime.

As we see, only focusing on the quantity of biomass harvested is not enough to reflect the complexity of the farmer rationality. It strongly depends on strategic choices about mobilizing stocks of dry forage and the value given to the biomass at different times of the year. These choices reflect production maximization or cost saving objectives. They determine the management regimes of each field in interaction with the management regimes of the other years and other parcels. Ecologically, the issues of biodiversity conservation do not depend solely on production "sacrifices" made by the breeder but more generally on its rationality. Thus, management regimes corresponding to a minimization of production costs may generate good environmental performance without these being explicitly sought by the farmer. Therefore, recent trends of biodiversity in European agroecosystems seem to depend as much on a modification of agricultural practices by themselves as on more general shifts in farmer rationalities.

Interactions between Management Regimes within a Farming System

At farm scale the trade-off between production and conservation is made possible by the combination of a variety of management regimes (land uses). Each management regime has a different aim and it is their interactions within the farming system that define the trade-off between production and conservation. The concavity of the curve linking productive and ecological performances (Figure 6) illustrates the mechanisms of interaction on both productive and ecological aspects.

Beyond the mechanisms of interaction between mowing and grazing included in the model, complementarity (on the productive aspects) between grazing regimes is a major asset for the farmer to reduce its dependency on dry forage. Thus, the combination, in adequate proportions, of productive and ecological management regimes may limit the periods at which dry forage is required (Figure 9).

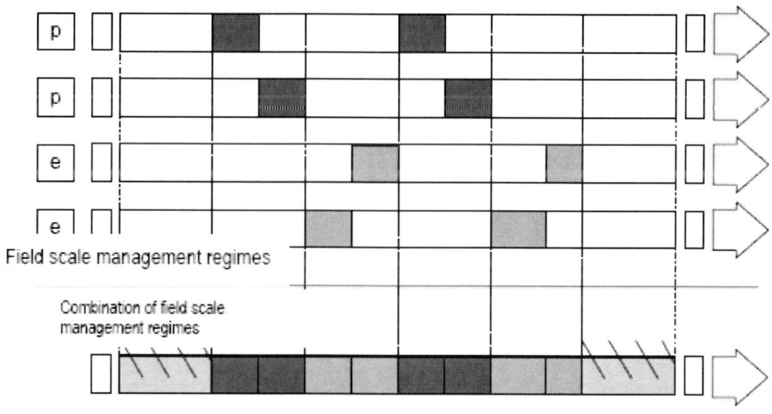

Figure 9. Complementarity (on the productive aspect) between a diversity of grazing management regimes. Management regimes p (dark grey) corresponds to a productive management regime, management regimes e (light grey) correspond to ecological ones. In the sequence combining the two modes of operation, hatched area correspond to periods of dry forage intake.

Complex Spatial Structure of Landscapes Facilitates Mechanisms of Ecological Complementarity

The spatial arrangement of fields makes it possible to overcome the trade-off between agricultural production and biodiversity conservation even in landscapes shaped by very productive objectives. For the same proportion of management regimes, a complex structure of the landscape favours mechanisms of ecological complementarity (Dunning *et al.,* 1992) (Figure 10).

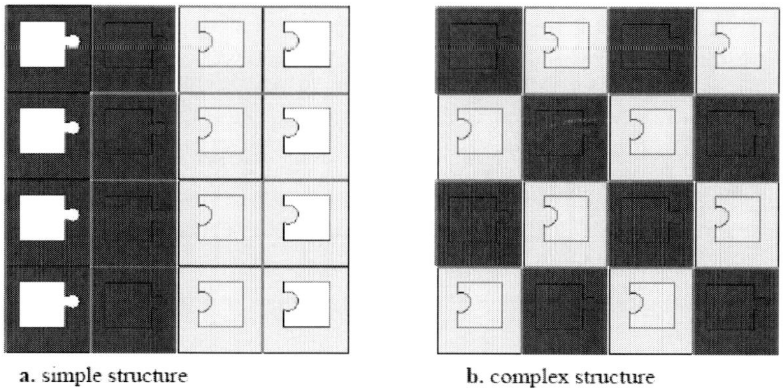

Figure 10. Examples of simple (a) and (b) complex structure within an agricultural landscape and importance of mechanisms of complementarity. Plots occupied by two complementary management regimes are shown (dark and light grey). Plots with no contact area with a complementary plot are represented with a white puzzle piece; those in contact with a complementary plot are represented with a coloured puzzle piece.

Here, it is the spatial diversity of management regimes that promotes the reconciliation between production and conservation. In terms of agricultural production, the spatial arrangement of agricultural fields is a major constraint in the organization of the farm territory. In particular, the geographical proximity of the fields is a strong driver of the spatial allocation of interacting management regimes. The spatially explicit model presented here does not include this dimension and the inclusion of spatial allocation rules of land uses is a priority perspective of this work.

Toward Landscape-Scale Management Schemes

Spatial structure of the landscape could be an efficient management tool for sustainable management of agricultural areas. Conservation policies at landscape scale would imply coordination between different stakeholders. This raises the question of collective management. Coordination among farmers is efficient as soon as farmers find a common interest (e.g. Hannachi *et al.,* 2009) but the question here is whether farmers would be willing to coordinate themselves on biodiversity issues. Only a few examples of landscape scale management have been developed and even fewer studies have been conducted on these schemes. However the example of Corn Crake (*Crex crex*) management in the lower Angevin valley in France showed quite positive results. On this example, biodiversity issues could federate farmers as soon as they acquired the status of experts (Le Guen and Sigwalt, 1999) or biodiversity became a way of valorising agricultural productions (Billaudeau and Thareau, 2008).

The set up of structures such as the Dutch environmental cooperatives (Renting andVan der Ploeg 2001) could be a good way of revealing common interests related to biodiversity. In such a context, models such as our can be used as catalysts of collective management as they provide tools to construct a shared point of view within the different actors.

Limits and Perspectives

The main limits of this work concern the accuracy of the quantitative predictions of the models. The first set of limits concerns the assessment of performances on the agricultural point of view. In this work the indicator of productive production is the harvested biomass. This indicator is strongly correlated to the feeding cost and gives a first view of the performance of the system. However, it gives no information on the primary production of the system (meat in this case). The use of an indicator such as the Autonomous Meat Production developed by Jouven (2006) could be a major improvement in this direction. The second set of limits concerns the ecological dynamics of the system. The three models developed here mainly aim at studying the interactions between productive and ecological dynamics, therefore the main efforts were put on the links between the two dimensions of the system. However some functions such as the movement function remain quite theoretical. First observations conducted on the field study area however are in line with this theoretical function (Lemel, 2007).

In spite of these limitations that make it impossible to give accurate quantitative predictions, the four main qualitative results found here are in line with general results given by previous studies. (1) The best ecological performances are reached at intermediate levels of productive performance (e.g. Uematsu *et al.* 2010, Marriott *et al.*, 2009), (2) result-oriented schemes improve the reconciliation between ecological and productive performances of an agroecosystem (Schwarz *et al.* 2008), (3) the performance of a landscape is not the average performance of its different land uses, and interactions between different land uses are major drivers of the trade-off between production and conservation (Vandermeer and Perfecto, 2007; Perfecto and Vandermeer, 2010), (4) the spatial structure of a landscape makes it possible to overcome the trade-off between productive and ecological performances (Polasky *et al.*, 2005, Groot *et al.*, 2007).

CONCLUSION

The different grassland management practices are not independent and the reconciliation between production and conservation depends on their articulation within management regimes that interact both spatially and temporally. Starting from a situation at field scale where only a policy aiming at saving production costs can reconcile production and conservation, moving to each of the larger scales increases the leeway of the farmer. Each level brings new possibilities of interaction between management regimes, and thus new levers for conciliation between production and conservation. The interaction between a diversity of management regimes taken as a whole is the basis of trade-off mechanisms between production and conservation.

REFERENCES

Altieri, M., and Nicholls, C. I. (2005). *Agroecology and the search for a trully sustainable agriculture.* Berkeley, CA: University of California.

Aubin, J.-P. (1991) *Viability theory.* Boston, MA: Birkäuser.

Beintema, A. J., G.J.D.M, Muskens (1987). Nesting success of birds breeding in Dutch agricultural grassland. *Journal of Applied Ecology, 24,* 743-758.

Beranger, C., and Micol, D. (1981). Use of grass by cattle at pasture. Importance of stocking rate and grazing system. *Fourrages, 122* 73-93.

Billaudeau, V., and Thareau, B. (2008). "L'éleveur et l'Oiseau": territoires de communication. *Colloque Economie sociale et solidaire : Nouvelles pratiques et dynamiques territoriales*, Nantes.

Brussard, L., Caron, P., Campbell, B., Lipper, L., Mainka, S., Rabbinge, R., Babin, D., and Pulleman, M. (2010). Reconciling biodiversity conservation and food security: scientific challenges for a new agriculture. *Current Opinion in Environmental Sustainability, 2*, 34-42.

Caswell, H. (1989). *Matrix population models* (second edition), Sunderland, MA: Sinauer Associates.

Clark, W. C., and Mangel, M. (2000). Dynamic state variable models in ecology: Methods and applications. Oxford: Oxford University Press.

De Lara, M., and Doyen, L. (2008). *Sustainable management of natural resources,* Berlin: Springer.

Dunning, J. B., Danielson, B. J., and Pulliam, H. R. (1992). Ecological processes that affect populations in complex landscapes. *Oikos, 65,* 169-175.

Durant, D., Tichit, M., Fritz, H., and Kerneis, E. (2008a). Field occupancy by breeding lapwings Vanellus vanellus and redshanks Tringa totanus in agricultural wet grasslands. *Agriculture Ecosystems and Environment, 128,* 146-150.

Durant, D., Tichit, M., Kerneis, E., and Fritz, H. (2008b). Management of agricultural wet grasslands for breeding waders: integrating ecological and livestock system perspectives - a review. *Biodiversity and Conservation, 17,* 2275-2295.

Flint, V. (1998) Waders as indicators of biological diversity. *International Wader Studies, 10,* 23.

Gibon, A., Sheeren, D., Monteil, C., Ladet, S., Balent, G. (2010). Modelling and simulating change in reforesting mountain landscapes using a social-ecological framework. *Landscape Ecology 25,* 267-285.

Green, R. E., Cornell, S. J., Scharlemann, J. P. W., and Balmford, A. (2005). Farming and the fate of wild nature. *Science, 307,* 550-555.

Groot, J. C. J., Rossing, W. A. H., Jellema, A., Stobbelaar, D. J., Renting, H., and Van Ittersum, M. K. (2007). Exploring multi-scale trade-offs between nature conservation, agricultural profits and landscape quality--A methodology to support discussions on land-use perspectives. *Agriculture Ecosystems and Environment, 120,* 58-69.

Guérin G., Léger F. and Pflimlin A. (1994). *Stratégies d'alimentation. Méthodologie d'analyse et de diagnostic de l'utilisation et de la gestion des surfaces fourragères et pastorales,* Paris : Institut de l'Elevage.

Hannachi, M., Coleno, F. C., and Assens, C. (2009). Collective strategies and coordination for the management of coexistence: the case studies of Alsace and western South of France. *4th International conference on coexistence between genetically modified (GM) and non-GM based agricultural supply chains, pp. 13.* Melbourne, Australia.

Jouven, M. (2006). Quels équilibres entre production animale et utilisation durable des prairies dans les systèmes bovins allaitants herbagers ? Une approche par modélisation des interactions conduite-troupeau-ressources, PhD, INA P-G.

Le Guen, R., and Sigwalt, A. (1999). Le métier d'éleveur face à une politique de conservation de la bio-diversité. *Economie Rurale, 249,* 41-48.

Lemel, J. Y. (2007). Système d'élevage performants dur le plan de la conservation des limicoles, Rapport final. Paris : INRA.

Marriott, C. A., Hood, K., Fisher, J. M., and Pakeman, R. J. (2009). Long-term impacts of extensive grazing and abandonment on the species composition, richness, diversity and productivity of agricultural grassland. *Agriculture Ecosystems and Environment, 134,* 190-200.

Milsom, T. P., Langton, S. D., Parkin, W. K., Peel, S., Bishop, J. D., Hart, J. D., and Moore, N. P. (2000). Habitat models of bird species' distribution: an aid to the management of coastal grazing marshes. *Journal of Applied Ecology, 37,* 706-727.

Perfecto, I., and Vandermeer, J. (2010). The agroecological matrix as alternative to the land-sparing/agriculture intensification model. *Proceedings of the National Academy of Sciences of the United States of America, 107,* 5786-5791.

Polasky, S., Nelson, E., Lonsdorf, E., Fackler, P., and Starfield, A. (2005). Conserving species in a working landscape: Land use with biological and economic objectives. *Ecological Applications, 15,* 1387-1401.

Redfern, C. P. F. (1982). Lapwing nest sites and chick mobility in relation to habitat. *Bird Study, 29,* 201-208.

Renting, H., and Van der Ploeg, J. D. (2001). Reconnecting nature, farming and society: environmental cooperatives in the Netherlands as institutionnal arrangements for creating coherence. *Journal of Environmental Policy Planning, 3,* 85-101.

Sabatier, R (2010). Arbitrages multi-echelles entre production agricole et conservation de la biodiversité dans un agroecosysteme prairial, PhD, AgroParisTech, Paris, 228 p.

Sabatier, R., Doyen, L., and Tichit, M. (2010). Modelling trade-offs between livestock grazing and wader conservation in a grassland agroecosystem. *Ecological Modelling, 221,* 1292-1300.

Sabatier, R., Kerneis, E., and Tichit, M. (2008). Designing grassland landscapes for economic or ecological priorities: application to livestock farming and birds. *Empowerment of the rural actors: a renewal of farming systems perspectives: 8th European IFSA Symposium, Clermont-Ferrand, France, 6-10 July 2008.* , pp. 581-589. INRA SAD.

Schwarz, G., Moxey, A., McCracken, D. I., Huband, S., and Cummins, R. (2008). An analysis of the potential effectiveness of a Payment-by-Results approach to the delivery of environmental public goods and services supplied by Agri-Environment Schemes. L. U. P. Group, pp. 108.

Uematsu, Y., Koga, T., Mitsuhashi, H., and Ushimaru, (2010). Abandonment and intensified use of agricultural land decrease habitats of rare herbs in semi-natural grasslands. *Agriculture Ecosystems and Environment, 135,* 304-309.

Vandermeer, J., and Perfecto, I. (2007). The agricultural matrix and a future paradigm for conservation. *Conservation Biology, 21,* 274-277.

Vickery, J. A., Tallowin, J. R., Feber, R. E., Asteraki, E. J., Atkinson, P. W., Fuller, R. J., and Brown, V. K. (2001). The management of lowland neutral grasslands in Britain: effects of agricultural practices on birds and their food resources. *Journal of Applied Ecology, 38,* 647-664.

In: Grazing Ecology: Vegetation and Soil Impact
Editor: José Ramón Arévalo

ISBN 978-1-62100-436-3

Chapter 9

ASSESSMENT OF SOIL PHYSICAL PROPERTIES IN INTEGRATED CROP-LIVESTOCK SYSTEMS IN MOLLISOLS OF ARGENTINA

***Patricia L. Fernández*[1], *Carina R. Alvarez*[1] *and Miguel A. Taboada*[1,2,3*]**

[1]Facultad de Agronomía, Universidad de Buenos Aires.
Ciudad Autónoma de Buenos Aires, Argentina
[2]Instituto de Suelos, CIRN, INTA. Los Reseros y Las Cabañas S/N,
(1688) Hurlingham, Provincia de Buenos Aires, Argentina.
[3]CONICET (C1033AAJ)
CiudadAutónoma de Buenos Aires, Argentina

ABSTRACT

The growth of the world population demands more food production. This raises the need to conserve the natural resources and the environment. In the search to achieve this goal and with the emergence of new technologies, the production in Argentina has changed accordingly. Most of the croplands in Argentina are currently managed under continuous cropping (CC), which is based on the extensive use of zero-till (ZT) farming, application of herbicides and prevalence of soybean [*Glycine max* (L.) Merr.] within crop rotations. Although this production system is economically sound, at least on a short-term basis, it produces few agricultural products. The integration of crop and livestock (ICL) production is an interesting alternative to increase the buffering capacity of the agro-ecosystem against variations in the weather and the economy. In addition, ICL adds diversity to the entire production system. In this chapter, we describe a regional study was conducted in the northern Pampas of Argentina in order to compare soil quality and condition at proximal cropland sites managed under CC and ICL systems under ZT, in representative soils of the region (Typic Argiudoll *vs.* Typic Hapludoll). It is well known

* Email: fpl@agro.uba.ar. Address: Av. San Martín 4453 (C1417DSE) Ciudad Autónoma de Buenos Aires, Argentina.

that cattle trampling causes damage over the soil physical properties. The natural recovery of the soil physical properties comprises natural and anthropogenic forces, representing the balance of soil amelioration and degradation processes. Natural soil amelioration processes include wetting-drying cycles, swelling-shrinkage cycles, whereas anthropogenic forces include passage of vehicles, tillage, and animal trampling. The soil physical condition naturally recovers to the greatest extent when animals are completely excluded from the plot. In this work, the recovery processes were evaluated in an ICL system, by measuring the dynamics of the soil physical properties over four years. Treatments consisted of closure and grazing plots, in soils with contrasting superficial textures. Introduction of livestock into croplands can lead to shallow soil compaction, which is caused by the collapse of macropores and the deformation of soil structure. Soil bulk density (BD) is usually increased by grazing, as shown in the occurrence of classical, shallow compaction processes. Grazing-induced decreases in BD are less common and, when they do occur, they are not easily explained. They have been ascribed to the intensive transit of animals on saturated soil. Thus, it has addressed the relationship between BD and soil water content (θ) in winter grazing of crop residues in a ZT system. In short, we analyzed the effect of cattle trampling at different scales and over time, and particularly considered the decrease in BD after grazing.

Keywords: Grazing management, livestock impact, long term effects, soil conservation.

INTRODUCTION

The world's most fertile soils, which are found mainly in temperate regions under humid and subhumid climate (FAO, 1998; Hartemink, 2002), should respond in the coming decades to an increasing social pressure to meet food demand in a world of high population growth (Watson and Zakri, 2005). In the agricultural sector, many important topics arise in relation to the conservation of the natural resources and the environment. There is a need to harmonize land use with those functions related to the maintenance of ecosystem services (Viglizzo and Frank, 2006). Soil is regarded as a reservoir of germplasm "*in situ*", a refuge for pest regulators, weeds and pathogens, and responsible for atmospheric carbon sequestration, among others. To reach this harmonization, suitable technologies that allow the development of a region and countries by increasing productivity and no impact on sustainability are required (Satorre, 2004).

The Pampas region of Argentina, which is located in the south cone of South America, covers about 60 Mha and is part of one of the largest prairies of the world (Figure 1). This chapter reports the results obtained in the north of the Pampas region, the main cropland area of Argentina. This region is characterized by a temperate and humid climate. The highest rainfalls occur normally from December to March (summer), and the lowest ones in the winter period (Hall *et al.,* 1992). The predominant soils are Typic Argiudolls and Typic Hapludolls, which are developed from Aeolian (loess-like) sediments (Soriano *et al.,* 1991).

Traditionally, mixed systems have been the primary production in the Pampas region using conventional tillage (CT) (mould-board plough, disk, etc) systems, with lower yield crop respect the present productivity (Soriano *et al.,* 1991; Senigagliesi and Ferrari, 1993). Conventional tillage results in loss of soil organic matter (OM), a decrease in soil structural stability and an increase in soil erosion (Studdert *et al.,* 1997; García-Préchac *et al.,* 2004). Since the ´90s, intensified agricultural activity due to technological and economic reasons, combined with a simplification of the rotation, has displaced livestock grazing to areas with lower productive capacity (Senigagliesi and Ferrari, 1993; Díaz Zorita *et al.,* 2002; Bolliger *et al.,* 2006). Concomitantly, there has been a sharp increase in the area planted with soybean [*Glycine max* (L.) Merr.], with predominance of monoculture. This represents a risk in terms of sustainability, as it returns a low amount of residues with a low C/N rate to the soil. These characteristics favor a rapid decomposition of residues, which enhances the possibility of low soil coverage during almost the entire year (Sanford, 1982; Bathke and Blake, 1984).

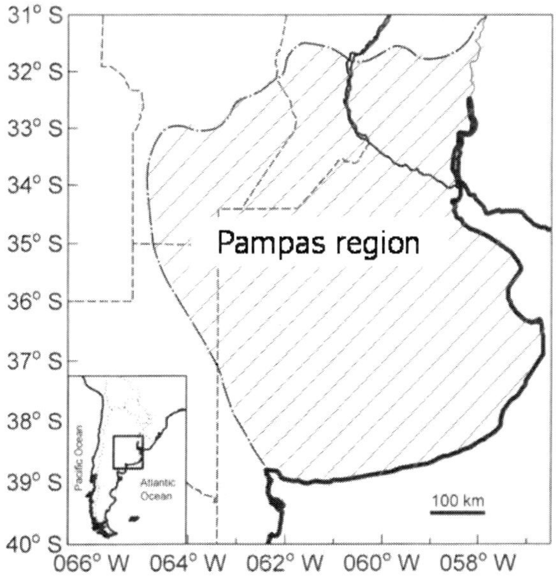

Figure 1. Map of the Pampas region located in Argentina.

Since the ´90 decade and due to different reasons (e.g. fuel saving, erosion control), the farmers of Argentina have increasingly adopted zero tillage (ZT) systems in replacement of CT systems. By 2007/08, about 85% of the agricultural area of Argentina was managed with ZT farming, being one of the countries with larger surface devoted to ZT farming, together with the United States and Brazil (Derpsch, 2011). The implementation of ZT is related to its numerous advantages, e.g. control of water and wind erosion, water conservation in the profile, mitigating the loss of OM and improvements in soil physical properties, saving operating time and reduction of fossil fuels (Díaz-Zorita *et al.,* 2002; Bolliger *et al.,* 2006; Steinbach and Alvarez, 2006; Lal *et al.,* 2007; Thomas *et al.,* 2007; Strudley *et al.,* 2008). The effectiveness of these systems depends largely on the type and amount of residues that will contribute to soil cover (Lal *et al.,* 2007).

Most of the cropland in the north of the Pampas region is managed under continuous cropping (CC), which is based on the extensive use of ZT farming, extensive application of herbicides (e.g. glyphosate and atrazine) to achieve weed-free fallow, and a scheme of succession of summer crops, mainly maize (*Zea mays* L.) followed by soybean. However, systems under ZT do not seem to improve the properties of some soils of this region (Andriulo and Cardone, 1998; Díaz-Zorita *et al.,* 2002; Constantini *et al.,* 2006; Sasal *et al.,*

2006; Micucci and Taboada, 2006; Alvarez *et al.,* 2009). Similar results have been found in other soils managed under ZT in South America (García-Prechac *et al.,* 2004, Bolliger *et al.,* 2006) and North America (Franzluebbers and Follett, 2005; Martens *et al.,* 2005). Two examples of these results are that the organic carbon (OC) in soils under ZT systems is usually only slightly higher than in the pre-existing conditions managed under CT (Lal, 2001; Díaz-Zorita *et al.,* 2002; Bolliger *et al.,* 2006; Micucci and Taboada, 2006), and that the compaction of soils under ZT (Thomas *et al.,* 1996; Taboada *et al.,* 1998 ; Díaz-Zorita *et al.,* 2002; Sasal *et al.,* 2006) presents an increase in BD and penetration resistance (PR) (Alvarez and Steinbach, 2009).

In recent years, the reintroduction of crop-livestock systems in ZT has been an interesting alternative, returning to the annual crop rotation with pastures. The integration of crop and livestock (ICL) systems not only allows increasing the buffering capacity of the agroecosystems against the weather and the economy (Viglizzo, 1986), but also adds diversity to the entire production systems. The inclusion of perennial pasture in the rotation restores the soil properties that were affected during the agricultural period (Campbell, 1978, Haynes *et al.,* 1991). This may contribute to increasing soil OC (Studdert *et al.,* 1997; Miglierina *et al.,* 2000; Gentile *et al.,* 2005) by greater production of aboveground and root biomass (Tisdall and Oades, 1982; Haynes *et al.,* 1991), which thus lead to an improvement of the soil structure and stability (Studdert *et al.,* 1997). Thus, the presence of pasture in the rotation, in combination with ZT, could entail net increases in soil OC, even showing a higher content of OC in CC under ZT (Siri-Prieto and Ernst, 2010). On the other hand, the presence of legumes in composite pastures would contribute to biologically fixed nitrogen. This path of N incorporation reduces costs and energy demand (Hoeppner *et al.,* 2006), and reduces the environmental impact. Research conducted at several locations in the humid Pampas region has shown that the contribution of N by biological fixation in alfalfa is in the order of 100 to 300 kg ha^{-1} yr^{-1} (Alvarez and Steinbach, 2010).

In summary, the advantages shown by ICL under ZT systems are: a- reduction of risks, like the inter-annual variations in the weather and the economy; b- greater diversity of the products offered; c- cost reduction in nitrogen fertilizers; d- potential increase in soil OC content. The production scheme of ICL under ZT systems includes consociated pastures (grass-legume) alternated with the grazing of crop residues and weeds during the winter (Beauchamp, 1990; Smil, 1999; Díaz-Zorita *et al.,* 2002; García-Préchac *et al.,* 2004; Franzluebbers and Stuedemann, 2008; Quiroga *et al.,* 2009). Crop residues and weeds represent a large source of biomass for ruminant feed and energy, since part of the forage supply can feed the animals, maintaining the quality of pastures in winter periods (Correa Luna, 2005). It reduces the use of herbicides during the fallow.

Cattle treading could lead to soil physical degradation (McCalla *et al.,* 1984; Holt *et al.,* 1996; Martínez and Zinck, 2004), in the form of soil compaction (Willatt and Pullar, 1983; McCalla *et al.,* 1984; Chanasyk and Naeth, 1995; Holt *et al.,* 1996; Greenwood *et al.,* 1997; Martínez and Zinck, 2004), and soil structural damage (Mapfumo *et al.,* 1999; Ferrero and Lipiec, 2000; Imhoff *et al.,* 2000; da Silva *et al.,* 2003). The consequence of cattle trampling depends on the grazing intensity (McCalla *et al.,* 1984; Chanasyk and Naeth, 1995), the stocking rate (McCalla *et al.,* 1984), the duration of the grazing period (Chanasyk and Naeth, 1995; Correa and Reichardt, 1995; Twerdoff *et al.,* 1999), and the animal species (Chanasyk and Naeth, 1995; Gijsman and Thomas, 1996; Mapfumo *et al.,* 1999; Rodd *et al.,* 1999; Taddese *et al.,* 2002). The soil response to treading depends on intrinsic factors, such as texture (McCalla *et al.,* 1984; Chanasyk and Naeth, 1995), vegetation species (Chanasyk and Naeth, 1995), soil water content (θ) (McCalla *et al.,* 1984; Chanasyk and Naeth, 1995; Smith *et al.,* 1997; da Silva *et al.,* 2003), and OM (Smith *et al.,* 1997; Quiroga *et al.,* 1999), among others. As a consequence, soil bulk density (BD) and penetration resistance (PR) often increase and the infiltration rate (IR) decreases with respect to ungrazed (UN) situations (Greenwood *et al.,* 1997, 1998; Singleton and Addison, 1999; da Silva, *et al.,* 2003; Franz-luebbers and Stuedemann, 2008).

The implementation of ICL in ZT systems in soils with high production capacity, based on maize and soybean, and grazing of crop residues poses a series of questions that require further studies. This chapter will summarize the results obtained in experiments conducted in the northern Pampas at a regional scale, comparing CC production systems with ICL systems, both under ZT, assessing changes in the physical properties and soil OC. We will also show the physical properties obtained from the analysis of several years of winter grazing compared to closures in ICL systems. In this sense, the regeneration of the soil with grazing exclusion and the temporal evolution of the PR over several years was assessed. Finally, a model explaining unanticipated changes in BD caused by winter grazing of residues under high θ was studied.

Regional Analysis of Production Systems

A regional analysis was performed with the aim to compare the soil physical condition and total OC (TOC) and its fractions between CC and ICL systems, both under ZT. Sampling included 22 productive situations (sampling units), involving two types of management (CC *vs.* ICL) and two soil textural groups (loam-sandy loam *vs.* silt loam). In the ICL systems, fields were alternatively cropped (alternating maize and soybeans, one per year) over 8 years, followed by a grass–alfalfa (lucerne, *Medicago sativa* L.) pasture for 4 years. In the CC systems, the crop sequence was maize followed by soybean. Plots selected for the ICL systems were sampled over mid-way through the agricultural cycle, i.e. 4 years after pasture (more detail in Fernández *et al.,* 2011).

Multivariate analysis was used to determine the principal gradients responsible for variability among sampling units. This analysis provides a summary of the information and identifies the most relevant data that can explain most of the variability within the dataset, determining an integration of variables denoted principal components (PC) (Figure 2). Soil texture (sand, clay and silt) and TOC and fractions of OC expressed as mass or concentration were variables that allowed separating sampling units with respect to the first axis. The first PC axis explained 47 % in the 0-5 cm layer and 48 % in the 5-20 cm layer of the variance. These results agree with those found by the analysis of variance (data not shown), which showed that ICL systems in the north of the Pampas region do not show higher contents of TOC nor of its fractions in the middle of the agricultural phase. Units sampling have been separated by TOC content and resistant OC (ROC), to be related to textural groups. The amount of fine mineral particles was one of the main factors controlling TOC and ROC levels (Fernández *et al.,* 2011). The second PC axis explained 25 % and 27 % of the variance for surface and subsuperficial layers, respectively; the variables that most significantly contributed were PR and θ, but appearing in opposite direction (Figure 2) (Fernández *et al.*, 2011).

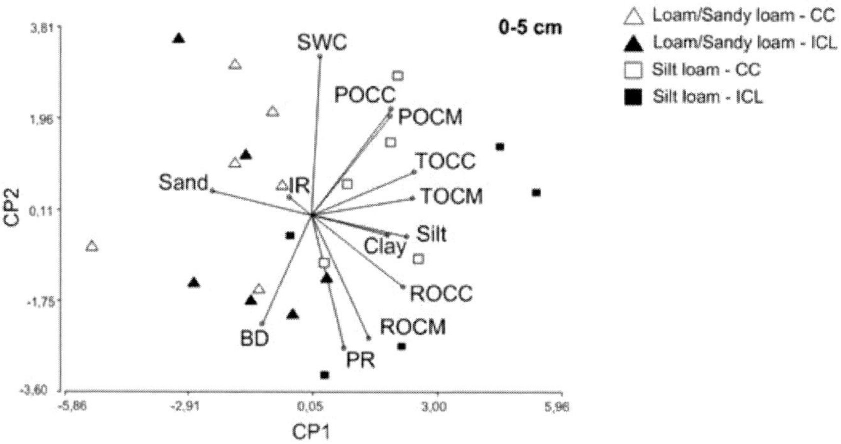

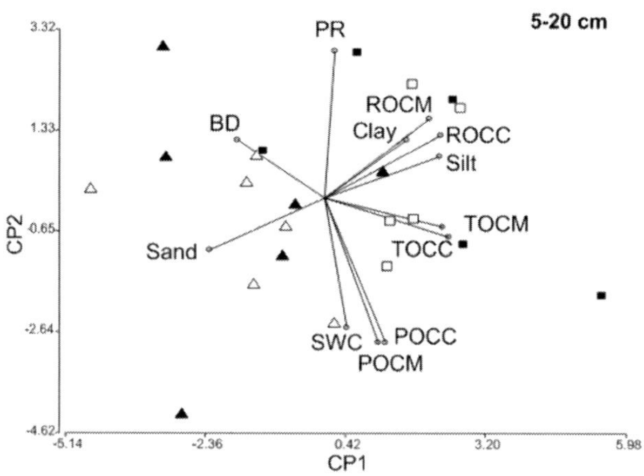

Variable	Axis 1	p-value	Axis 2	p-value
Sand	-0.35	***	0.09	
Clay	0.27	***	-0.08	
Silt	0.34	***	-0.09	
TOCC	0.38	***	0.11	
ROCC	0.33	***	-0.26	*
POCC	0.30	***	0.31	*
TOCM	0.37	***	0.02	
ROCM	0.26	**	-0.38	***
POCM	0.30	*	0.29	*
BD	-0.19		-0.34	**
SWC	0.05		0.51	***
PR	0.10		-0.44	***
IR	-0.08		0.06	

Variable	Axis 1	p-value	Axis 2	p-value
Sand	-0.36	***	-0.16	
Clay	0.26	**	0.18	
Silt	0.36	***	0.13	
TOCC	0.39	***	0.12	
ROCC	0.36	***	0.20	
POCC	0.19	*	-0.45	***
TOCM	0.37	***	-0.09	
ROCM	0.33	***	0.25	*
POCM	0.17		-0.45	***
BD	-0.27	***	0.18	
SWC	0.07		-0.40	***
PR	0.03		0.46	***

Figure 2. Biplot of principal component analysis. Vectors indicate the relative weight of each variable on the axes. *P<0.05; **P<0.01; ***P<0.001 (Fernández *et al.,* 2011). TOCC, total organic carbon concentration; ROCC, resistant organic carbon concentration; POCC, particulate organic carbon concentration; TOCM, total organic carbon mass; ROCM, resistant organic carbon mass; POCM, particulate organic carbon mass; BD, bulk density; SWC, soil water content; PR, penetration resistance.

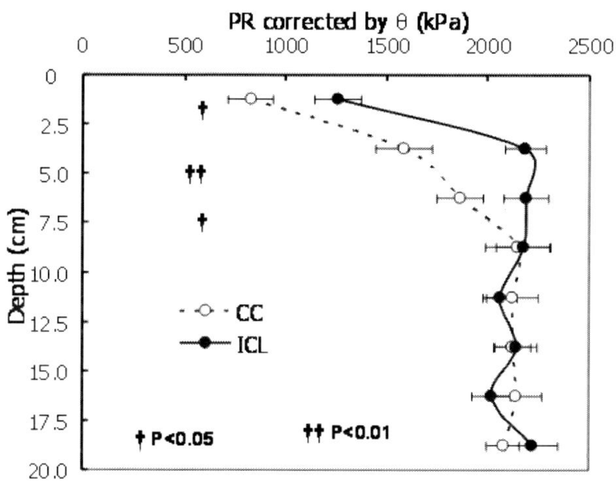

Figure 3. Soil penetration resistance (PR) corrected by gravimetric θ at depth, under CC (continuous cropping) and ICL (integrated crop–livestock) systems. θ mean at 0-5 cm: 22.7 %, θ mean at 5-20 cm: 20.4 %. Horizontal bars represent standard error (Fernández *et al.,* 2011).

In order to improve the interpretation of PR results, which can be masked by θ at the time of measurement, some authors (De León González *et al.,* 1998; da Silva et al. 2003; Franzluebbers and Stuedemann 2008) have noted the appropriateness of correcting field RP values by their θ. The slope of fitted lines in PR – θ regressions was used as a correction factor for PR. PR corrected by θ showed a significance differences between management up to a depth of 7.5 cm (Figure 3).

Increases in PR were not a consequence of increases in BD, showing that total porosity was not affected by grazing (Fernández *et al.,* 2011). These increases in PR cannot be attributed to compaction processes, which typically occur in parallel to soil densification

(Hamza and Anderson, 2005). Rather than compaction, the increase in soil PR found here is likely to be in response to a process of topsoil hardening, as previously found in ZT topsoil of the Pampas region (Taboada et al,. 1998; Díaz-Zorita *et al.,* 2002; Alvarez *et al.,* 2009). Taking these PR profiles into account, it can be thought that preexistent soil hardness under ZT farming is further promoted by grazing (Fernández *et al.,* 2011).

The critical threshold value of RP (2000 kPa) reported for penetrometer with a 30º tip angle (Gupta and Allmaras, 1987; Glinski and Lipiec, 1990) was only exceeded below a depth of 0.05 m in ICL systems. This may bring consequences for the driest years in the early stages of crops following the winter grazing. However, values above the threshold were found from 10 cm depth in both management systems.

Soil management practices could change soil macroporosity, being this change of great interest because of its influence on air capacity and soil water dynamics (Alakukku, 1998). The pore size distribution can be obtained by various methodologies that are characterized by their complexity. The determination of the shrinkage curve (ShC) of a soil is a simple methodology from which important soil parameters can be derived. After drying a single soil core, its resulting ShC allows assessing physical parameters and monitoring the changes that occur as a consequence of anthropic activity.

Soil shrinkage is defined as the specific volume (V) change of soil relative to its gravimetric θ (Haines, 1923; Stirk, 1954), or pore volume changes when the soil becomes drier (Braudeau *et al.,* 1999) (Figure 4).

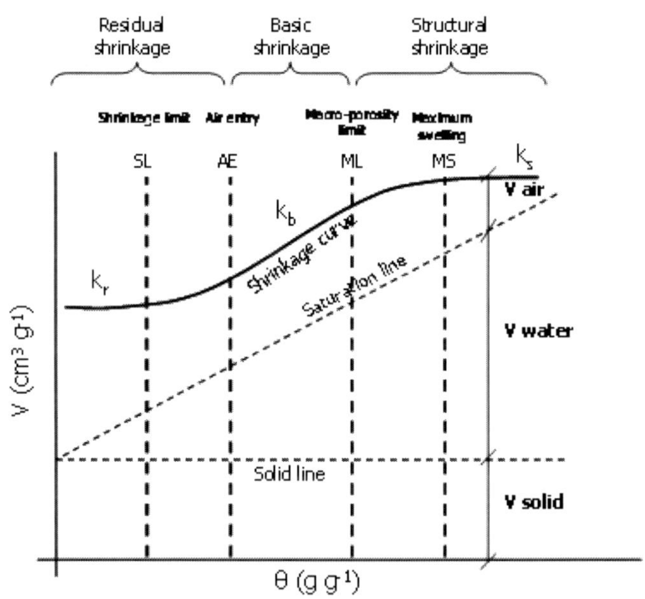

Figure 4. A soil shrinkage curve model, with the transition points: MS, maximum swelling; ML, macro-porosity limit; AE, air entry; SL, shrinkage limit; and linear slope: k_s, structural slope; k_b, basic slope; k_r, residual slope; Vsolid, solid specific volume; Vwater, water specific volume; Vair, air specific volume (adapted from Boivin *et al.,* 2006).

Inclusion of non-cropped enclosed areas of land around large country houses (quasi-pristine, QP) allows evaluating the degradation and soil behavior between the target systems (CC *vs.* ICL). One of the most important parameters describing soil shrinkage is the linear slope of the structural phase (k_s), which is intimately related to macro-pore stability. Data show that for the two soils tested (Typic Argiudoll- silt loam *vs.* Typic Hapludoll- sandy loam), QP management showed the lowest values of k_s, intermediate for ICL management, and the highest values for CC management (QP < ICL < CC) (Table 1). High k_s values indicate macro-pore collapse when the soil dries, determining the θ corresponding to the transition point ME (θ_{ME}) (greater than 10 μm) (Boivin *et al.,* 2004). The macro-pores V (V_{ma}) at the point of transition ME was higher in the QP management, followed by ICL, and finally by CC management (CP > IAG > AC), presenting an opposite tendency on the plasma porosity V (V_p) (Figure 5). This shows that there is a change in the distribution of pore sizes, along with an increase in the proportion of micro-pores at expense of a decrease in macro- and meso-pores.

Table 1. Shrinkage properties: slope of the structural shrinkage (k_s), slope of the basic shrinkage (k_b), water content and bulk specific volume coordinates of the transition points SL, AE, ML and MS determined with the XP model. Soil type: Typic Hapludoll and Typic Argiudoll. Management: Quasi-pristine (QP), integrated crop-livestock (ICL), continuous cropping (CC)

		Typic Hapludoll			Typic Argiudoll		
		QP	ICL	CC	QP	ICL	CC
Parameter		Shrinkage properties					
k_b	$cm^3\ g^{-1}$	0.116	0.069	0.067	0.174	0.099	0.112
k_s	$cm^3\ g^{-1}$	0.094	0.123	0.219	0.094	0.190	0.265
θ_{SL}	$g\ kg^{-1}$	37	40	65	34	44	68
V_{SL}	$cm^3\ g^{-1}$	0.735	0.654	0.664	0.912	0.719	0.662
θ_{AE}	$g\ kg^{-1}$	48	72	94	89	69	79
V_{AE}	$cm^3\ g^{-1}$	0.737	0.655	0.665	0.917	0.721	0.663
θ_{ML}	$g\ kg^{-1}$	105	165	180	189	138	187
V_{ML}	$cm^3\ g^{-1}$	0.749	0.661	0.670	0.934	0.728	0.675
θ_{MS}	$g\ kg^{-1}$	124	205	294	316	260	238
V_{MS}	$cm^3\ g^{-1}$	0.752	0.665	0.688	0.950	0.747	0.686

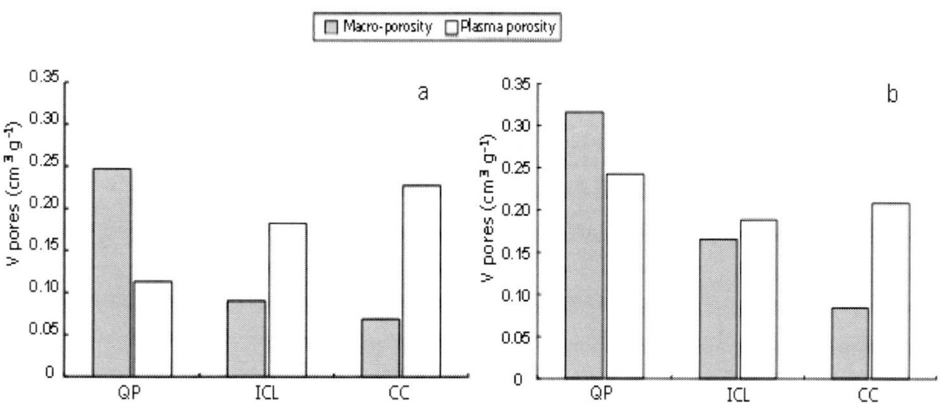

Figure 5. Specific volume of macroporosity and plasma porosity for management: quasi-pristine (QP), integrated crop-livestock (ICL), continuous cropping (CC); in Typic Hapludoll (a); Typic Argiudoll (b).

The slope of the linear basic phase (k_b) is another important parameter. The shrinkage of this phase is related to the plasma component, where the type and proportion of clay determines the variations of V (Braudeau and Mohtar, 2006) together with aggregate stability (McGarry and Daniels, 1987). Boivin et al. (2006) noted that k_b values lower than 0.200 cm^3 g^{-1} correspond to compacted or rigid soils. Soils classified as rigid present almost constant V values as the soil dries. QP management, here used as a reference for its management and history characteristics, showed values lower than 0.200 cm^3 g^{-1}, which allows to inferring the rigid condition of these soils (Table 1).

Bulk density is a parameter that allows the determination of soil total porosity. However, it does not reveal the quality (i.e. connectivity, orientation, or size distribution) of the porosity. The results show that despite the absence of BD differences between managements, the macro-porosity of QP was higher than that in ICL and CC managements, finding low values of CC compared with the ICL system.

Effects of Four Years of Winter Grazing over Soil Physical Properties

The ability of soils to support ICL under ZT systems can be evaluated from the regeneration capacity of soils. The harmful effects of livestock trampling on soil physical properties have been well documented (Drewry and Paton, 2000; Greenwood *et al.,* 1997, 1998; Greenwood and McKenzie, 2001). However, compaction could be alleviated through natural mechanisms (Drewry, 2006), generally little studied by researchers. The implementation of grazing exclosure has been reported as a necessary condition to assess the impact of cattle treading and soil amelioration, through improvements in soil physical condition (Greenwood and McKenzie, 2001; Zegwaard *et al.,* 1998; Singleton *et al.,* 2000). Natural recovery processes include wetting-drying cycles, growth and decomposition of roots and soil biological activity (Larson and Allmaras, 1971; Greenland, 1981; Grant and Dexter, 1989; Dexter, 1991; Greenwood and McKenzie, 2001). Compaction by trampling and the following natural recovery of soil physical properties show a cyclical pattern (Drewry *et al.,* 2004, Monaghan *et al.,* 2005) (Figure 6). This is related to the variation of rainfall across years in the study site.

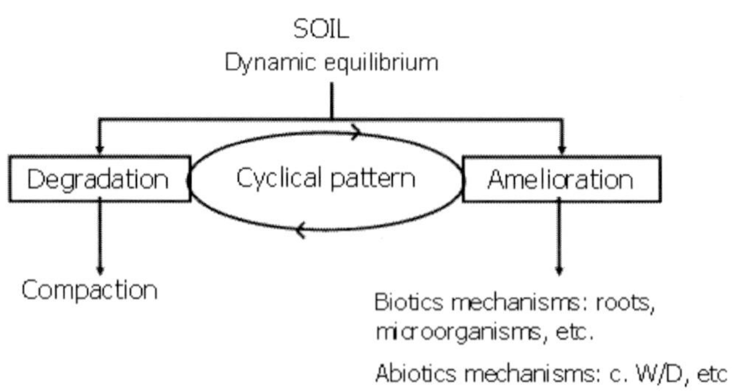

Figure 6. Conceptual model on the cyclical variation pattern of soil degradation by cattle trampling and amelioration processes in winter grazed crop-livestock systems.

Research was conducted in two representative soils (Typic Hapludoll and Typic Argiudoll) of the northern Pampas, over four years (2005 to 2008). Winter grazing of crop residues (treatment G) and closure plots (treatment UN) was evaluated during this period. Winter grazing showed no adverse changes in soil physical properties for both layers evaluated (Table 2). Probably, the absence of negative effects by trampling was related to a number of causes: a) the low stocking rate never higher than 1.1 cow ha^{-1} during this period, b) the higher soil bearing capacity of ZT soils, as compared to tilled soils; c) the usual low amount of rainfall during winter in the study area (Figure 7). As a result of these factors, compaction by trampling was prevented in this study.

In several studies, soil mechanical impedances have been identified as the cause of yield decreases for several crops in a wide range of soil textures (Arvidsson and Hakansson, 1996; Radford *et al.,* 2001; Tennant and Hall, 2001; Dauda and Samari, 2002; Hamza and Anderson, 2003, 2005; Sadras *et al.,* 2005). This is related to the difficulty of roots to explore and penetrate compacted horizons. However, in the study site there was no evidence of yield reduction after winter grazing, being consistent with the absence of physical modifications by grazing (Table 2).

Soil PR is a highly sensitive variable to identify and characterize the soil compacted layers (Martínez and Zinck, 2004). The PR evaluated over time provides information about the dynamics of change due to human activities determined by the production system. Figure 8 (a, b) shows the dynamics of the PR corrected by θ. The grazing periods showed no changes over the years, having the variation pattern an erratic behavior (Figure 8). However, PR decreased at final of crop cycle, and increased after the harvesting operation (Figure 8). Therefore, crop harvesting operations were the main stress factor for ICL systems. The occurrence of this stress takes place in autumn, a period with high rainfalls and, as a consequence, soils become more susceptible and unfavorable physical conditions appear. The presence of roots crop was the factor that led to soil loosening (Figure 8). Alakukku (1998) observed that roots crop can be a potential tool for soil compaction alleviation.

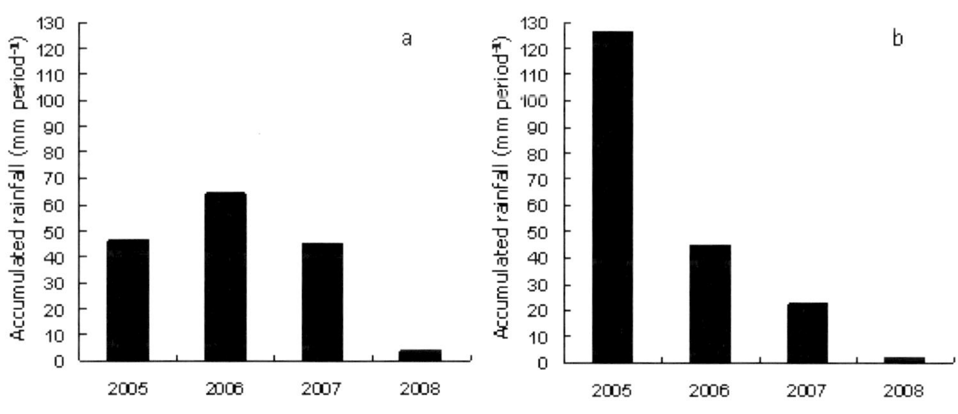

Figure 7. Accumulated rainfall for each grazing winter and year under study. Typic Hapludoll (a); Typic Argiudoll (b).

Table 2. Soil physical properties and crop yields for ungrazed (UN) and grazed (G) treatments, in Typic Hapludoll and Typic Argiudoll, for years: 2005, 2006, 2007, and 2008. Statistical analyses of variance, P-value < 0.05 was significant. BD, bulk density; SI, structural instability; IR infiltration rate. Standard error is presented between parentheses. N/D, no data

Typic Hapludoll													
		2005			2006			2007			2008		
	Units	UN	G	P-value	UN	G	P-value	UN	G	P-value	UN	G	P-value
BD (0-5 cm)	Mg m^{-3}	1.40 (0.02)	1.39 (0.01)	0.75	1.41 (0.02)	1.42 (0.02)	0.93	1.42 (0.02)	1.41 (0.02)	0.71	1.33 (0.02)	1.32 (0.03)	0.85
BD (5-20 cm)	Mg m^{-3}	1.45 (0.02)	1.44 (0.01)	0.67	1.46 (0.02)	1.40 (0.03)	0.38	1.50 (0.01)	1.46 (0.02)	0.30	1.41 (0.03)	1.40 (0.04)	0.89
SI (0-20 cm)	mm	0.59 (0.13)	0.61 (0.14)	0.93	0.54 (0.04)	0.53 (0.07)	0.93	0.29 (0.02)	0.47 (0.09)	0.15	0.79 (0.09)	0.76 (0.08)	0.79
IR	mm h^{-1}	48.01 (23.74)	53.94 (30.22)	0.88	76.05 (18.21)	66.68 (7.95)	0.66	43.78 (23.91)	62.23 (33.52)	0.68	42.18 (6.23)	40.98 (10.54)	0.93
		Maize			Soybean			Maize					
Crop yield	Mg ha^{-1}	10.75 (0.22)	12.01 (0.81)	0.21	3.44 (0.10)	3.46 (0.22)	0.92	N/D	N/D	-	N/D	N/D	-
Typic Argiudoll													
		2005			2006			2007			2008		
		UN	G	P-value	UN	G	P-value	UN	G	P-value	UN	G	P-value
BD (0-5 cm)	Mg m^{-3}	1.33 (0.04)	1.16 (0.04)	0.05	1.36 (0.06)	1.37 (0.04)	0.77	1.39 (0.05)	1.40 (0.04)	0.78	1.37 (0.04)	1.39 (0.03)	0.20
BD (5-20 cm)	Mg m^{-3}	1.30 (0.05)	1.26 (0.03)	0.43	1.46 (0.01)	1.45 (0.01)	1.00	1.38 (0.02)	1.45 (0.01)	0.02	1.43 (0.02)	1.42 (0.02)	0.68
SI (0-20 cm	mm	0.62 (0.01)	0.99 (0.10)	0.02	0.27 (0.03)	0.29 (0.03)	0.71	0.48 (0.04)	0.41 (0.02)	0.20	1.01 (0.09)	1.25 (0.07)	0.09
IR	mm h^{-1}	138.64 (50.52)	62.68 (23.71)	0.24	136.14 (44.62)	150.54 (45.64)	0.83	50.27 (13.38)	25.65 (5.88)	0.17	48.41 (12.88)	67.44 (28.61)	0.58
		Maize			Soybean			Maize					
Crop yield	Mg ha^{-1}	10.49 (0.22)	9.96 (0.26)	0.19	3.50 (0.09)	3.51 (0.22)	0.97	11.87 (0.34)	12.66 (0.24)	0.12	N/D	N/D	-

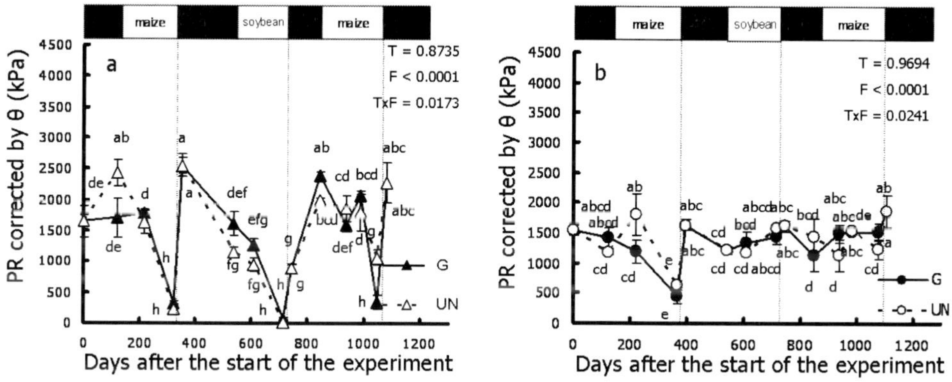

Figure 8. Soil PR corrected by θ variation with time for the 0-5 cm layer, Typic Hapludoll (a); and Typic Argiudoll (b). Black boxes indicate the grazing periods, white boxes indicate the cropping periods, and dotted lines indicate harvest operation. Ungrazed (UN) and grazed (G) treatments were analyzed as repeated measures over time. The covariance structure VC resulted in the best agreement with the Akaike information criterion (AIC; Akaike, 1974).

Analysis of Soil Bulk Density Decreases after Grazing

A conceptual model was developed to explain the causes of the unexpected changes in BD after winter grazing of crop residues under wet soil conditions (Fernández *et al.,* 2010). The research developed in this work showed that in the first year of sampling (2005) no compaction occurred as a result of grazing, but instead there was a lower BD in 0-5 cm in the G treatment in Typic Argiudoll (Table 2). That year coincided with high rainfall during winter grazing (Figure 5b), and moderate stocking (1.1 cow ha^{-1}). This unexpected behavior in the BD led us to explore the possible causes. Figure 9 describes different possible hypotheses, which could explain the reduction in BD after grazing. The first hypothesis was that the change in soil volume could be due to the classical development of soil swelling. Soil swelling requires the presence of high amounts of expansive clays. However, this hypothesis was not accepted because illite was the dominant clay in the region. Moreover, the phenomena of classical swelling require that the increase in soil θ leads to the replacement of air pores by water-filled pores (WFP ~ 100%). This replacement was not observed, since a significant air-filled porosity was observed at wet soil conditions (Fernández *et al.,* 2010).

Another mechanism explaining why soil volume changed in the G treatment was the occurrence of trapped air unable to bubble from soil pores. The thin section of this soil showed the development of a layered structure showing pores with horizontal orientation (Figure 10). Similarly, other authors in the region observed laminar structures in soil surfaces under ZT (Barbosa *et al.,* 1997; Bonel *et al.,* 2005; Sasal *et al.,* 2006). This type of structure does not favor the air-water exchange between the soil and the atmosphere. Therefore, this structure could affect the exchange dynamics due to lack of connectivity.

Using a laboratory device, Grant and Dexter (1990) generated high and low air pressure within soils (Figure 11). When the air pressure was high, the entry of water led to soil slaking, creating fissures allowing high infiltration processes. In contrast, soil with low air pressure led to soil swelling and slower infiltration. In 2005, IR showed lower values for the G treatment than for the UN treatment, although the difference was not significant (Table 2). Therefore, the evidence shows that lower BD values in G were due to trapped air due to lack of connectivity of the pores with the atmosphere, a mechanism exacerbated by livestock trampling in high soil θ. On the other hand, based on the evidence shown, air entrapment occurred at low pressures.

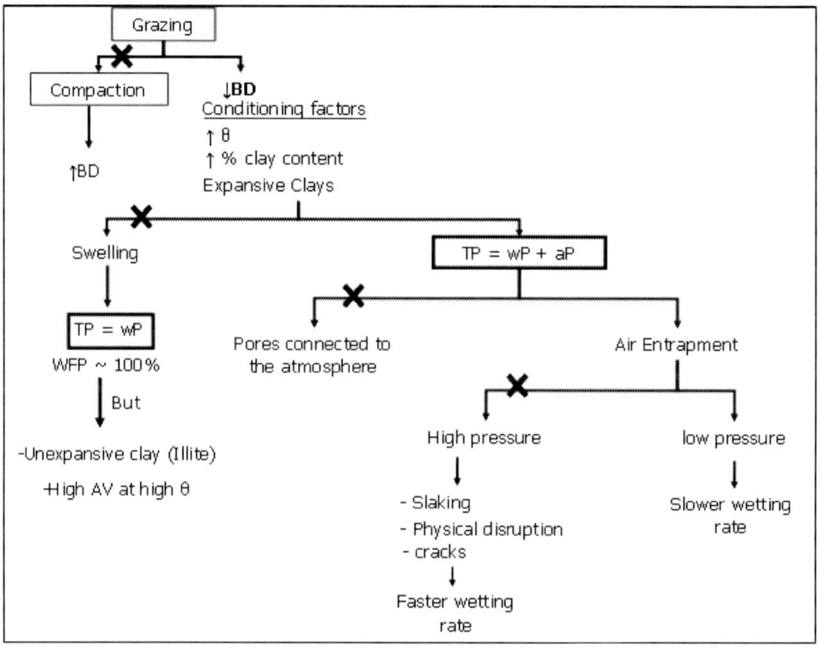

Figure 9. A model developed to understand soil swelling. TP, total porosity; wP, pores with water; WFP, water-filled porosity; AV, air volume; aP, pores with air.

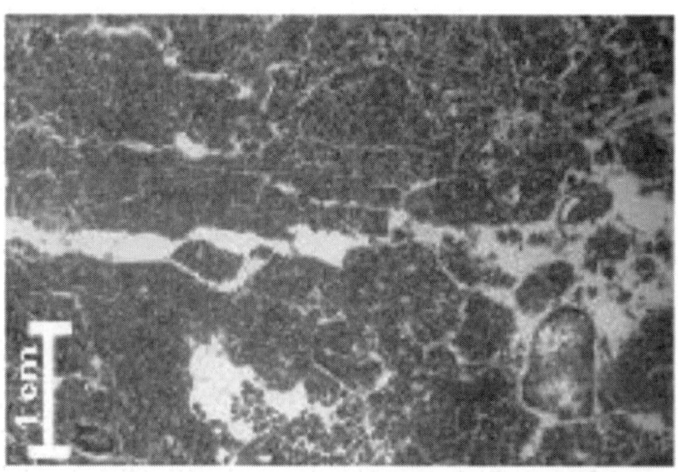

Figure 10. Thin section of superficial layer in the ICL system in the Typic Argiudoll.

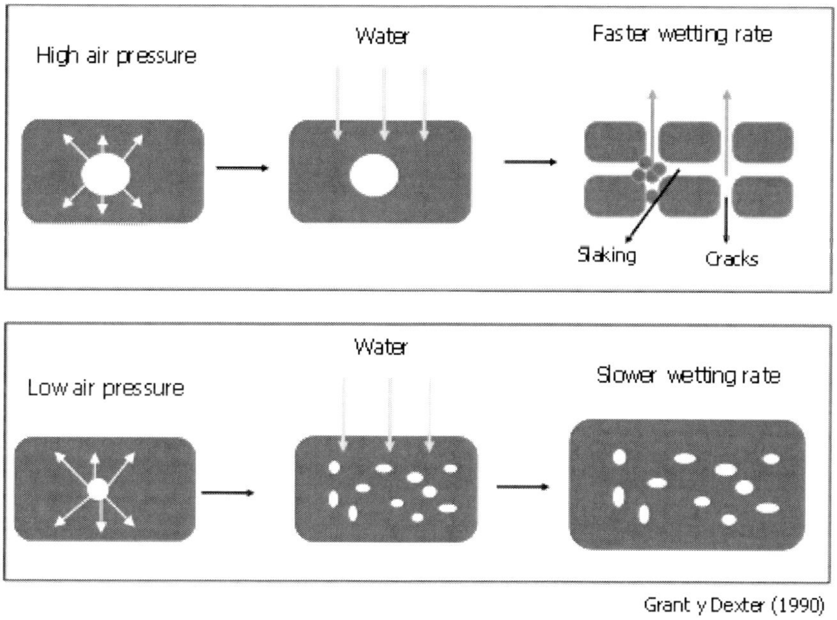

Figure 11. A model of high and low air pressure in the soil (adapted from Grant and Dexter, 1990).

CONCLUSION

Inclusion of pastures into the ICL systems did not show an increase in OC with respect to CC, both under ZT. Cattle grazing and trampling did not lead to compaction processes in the region studied. There was some slight topsoil hardening in ICL systems which exceeded the critical threshold for soil PR above 5 cm depth. Winter grazing of crop residues did not lead to changes in the soil physical properties due, mainly, to the low stocking rate and the low rainfall in this period. In 2005, a decrease in BD after the grazing period was analyzed. This soil behavior was explained by low air pressure entrapment, because grazing was at high soil water conditions and predominant horizontal soil pores with low connectivity.

REFERENCES

Akaike, H. (1974) A new look at the statistical model identification. IEEE *Trans. Automatic Control AC-19*, 716-723.

Alakukku, L. (1998) Properties of compacted fine-textured soils as affected by crop rotation and reduced tillage. *Soil and Tillage Research, 47,* 83-89.

Alvarez, C. R., Taboada, M. A., Gutierrez Boem, F., Bono, A., Fernández, P. L., and Prystupa, P. (2009) Topsoil properties as affected by tillage systems in the Rolling Pampa Region of Argentina. *Soil Science Society of America Journal, 73,* 1242-1250.

Alvarez, R., and Steinbach H. S. (2009) A review of the effects of tillage Systems on some soil physical properties, water content, nitrate availability and crops yield in the Argentine Pampas. *Soil and Tillage Research, 104*, 1-15.

Alvarez, R., and Steinbach H. S. (2010) Ciclado de nitrógeno en agrosistemas. In R. Alvarez, G. Rubio, C. R. Alvarez, and R. S. Lavado (Eds.), *Fertilidad de suelos: Caracterización y manejo en la Región Pampeana* (First edition, pp. 231-269). Buenos Aires, Argentina: Facultad de Agronomía Editorial.

Andriulo, A., and Cordone, G. (1998) Impacto de labranzas y rotaciones sobre la materia orgánica de los suelos de la región pampeana húmeda. In J. L. Panigatti, H. Marelli, D. Buschiazzo, R. Gil (Eds.), *Siembra Directa* (pp. 65-96) Buenos Aires, Argentina.

Arvidsson, J., and Håkansson, I. (1996) Do effects of soil compaction persist after ploughing? Results from 21 long-term field experiments in Sweden. *Soil and Tillage Research, 39,* 175-197.

Barbosa, A. O., Taboada, M. A., Rodríguez, M. B., and Cosentino, D. J. (1997) Regeneración de la estructura en diferentes fases de degradación de un suelo franco limoso de La Pampa Ondulada (Argentina). *Ciencia del Suelo, 15,* 81-86.

Bathke, G. R., and Blake, G. R. (1984) Effects of soybeans on soil properties related to soil erodability. *Soil Science Society of America Journal, 48,* 1398-1401.

Beauchamp, E. G. (1990) Animals and soil sustainability. *Journal of Agricultural and Environmental Ethics, 3,* 89-98.

Boivin, P., Gratier, P., and Tessier, D. (2004) Relationship between clay content type, and shrinkage properties of soil samples. *Soil Science Society of America Journal, 68,* 1145-1153.

Boivin, P., Schäffer, B., Temgoua, E., Gartier, M., and Steinman, G. (2006) Assessment of soil compaction using soil shrinkage modeling: experimental data and perspectives. *Soil and Tillage Research, 88,* 65-79.

Bolliger, A., Magid, J., Carneiro Amado, T. J., Skóra Neto, F., Dos Santos Ribeiro, M. de F., Calegari, A., Ralisch, R., and De Neergaard, A. (2006) Taking stock of the Brazilian "zero-till revolution": a review of landmark research and farmers' practice. *Advances in Agronomy, 91,* 47-109.

Bonel, B. A., Morrás, H. J. M., and Bisaro, V. (2005) Microstructure and organic matter modifications in an Argiudoll under different conditions of cultivation and conservation. *Ciencia del Suelo, 23,* 1-12.

Braudeau, E., Costantini, J. M., Bellier, G., and Colleuille, H. (1999) New device and method for soil shrinkage curve measurement and characterization. *Soil Science Society of America Journal, 63,* 525-535.

Braudeau, E., and Mohtar, R. H. (2006) Modeling the swelling curve for packed soil aggregates using the pedostructure concept. *Soil Science Society of America Journal, 70,* 494-502.

Campbell, C. A. (1978). Soil organic carbon, nitrogen and fertility. *Developments of Soil Science, 8,* 173-271.

Chanasyk, D., and Naeth, A. (1995) Grazing impacts on bulk density and soil strength in the foothills fescue grasslands of Alberta, Canada. *Canadian Journal of Soil Science, 75,* 551-557.

Constantini, A., De-Polli, H., Galarza, C., Pereyra Rossiello, R., and Romaniuk, R. (2006) Total and mineralization soil carbon as affected by tillage in the Argentinean Pampas. *Soil and Tillage Research, 88,* 274-278.

Correa, J. C., and Reichardt, K. (1995) Efeito do tempo de uso das pastagens sobre as propiedades de um Latossolo Amarelo da Amazonia Central. *Pesquisa Agropecuária Brasileira, 30,* 107-114.

Correa Luna, M. (2005) Cría Bovina Intensiva. Sistemas Ganaderos en Siembra Directa., AAPRESID. 11 y 12 de mayo de 2005, Rosario, Santa Fe (pp. 101-106).

da Silva, A. P., Imhoff, S., Corsi, M. (2003) Evaluation of soil compactions in an irrigated short-duration grazing system. *Soil and Tillage Research 70,* 83-90.

Dauda, A and Samari, A. (2002) Cowpea yield response to soil compaction under tractor traffic on a sandy loam soil in the semi-arid region of northern Nigeria. *Soil and Tillage Research, 68,* 17-22.

De León González, F., Payán Zelaya, F., and Sánchez, R. (1998) Localización de capas compactadas en el perfil del suelo mediante penetrometría. *Terra, 16,* 303-307.

Derpsch, R. (2011) Situación general de la siembra directa en el mundo. http://www.rolf-derpsch.com/. Acceso: 1/4/2011.

Dexter, A. R. (1991) Ameloration of soil by natural processes. *Soil and Tillage Research, 20,* 87-100.

Díaz-Zorita, M., Duarte, G., and Grove, J. (2002) A review of no-till systems and soil management for sustainable crop production in the sub-humid and semi-arid Pampas of Argentina. *Soil and Tillage Research, 65,* 1-18.

Drewry, J. J., and Paton, R. J. (2000) Effects of cattle treading and natural amelioration on soil physical properties and pasture under diary farming in Southland, New Zealand. *New Zealand Journal of Agricultural Research, 43,* 377-386.

Drewry, J. J., Paton, R. J., Monaghan, R. M. (2004) Soil compaction and recovery cycle on a Southland dairy farm: implications for soil monitoring. *Australian Journal of Soil Research, 42,* 851-856.

Drewry, J. J. (2006) Natural recovery of soil physical from treading damage pastoral soils in New Zealand and Australia: A review. *Agriculture, Ecosystems and Environment, 114,* 159–169.

Food and Agriculture Organization of the United Nations (FAO) (1998) Word Reference Base of Soil resources. (www.fao.org/docrep/W8594E/W8594E00.htm#Contents).

Franzluebbers, A. J., and Follett, R. F. (2005) Greenhouse gas contributions and mitigation potential in agricultural regions of North America: Introduction. *Soil and Tillage Research, 83,* 1-8.

Franzluebbers, A. J., and Stuedemann, J. A. (2008) Soil physical responses to cattle grazing cover crops under conventional and no tillage in the Southern Piedemont USA. *Soil and Tillage Research, 100,* 141-153.

Fernández, P. L., Alvarez, C. R., Schindler, V., Taboada, M. A. (2010) Changes in topsoil bulk density after grazing crop residues under no-till farming. *Geoderma, 159,* 24-30.

Fernández, P. L., Alvarez, C. R., and Taboada, M. A. (2011) Assessment of topsoil properties in integrated crop–livestock and continuous cropping systems under zero tillage. *Australian Journal of Agricultural Research, 49,* 143-151.

Ferrero, A., and Lipiec, J. (2000) Determining the effect of trampling on soils in hill slope-woodlands. *International Agrophysics, 14,* 9-16.

García-Prechác, F., Ernst Benech O., Siri-Prieto G., and Terra J. A. (2004) Integrating no-till into crop-pasture rotations in Uruguay. *Soil and Tillage Research, 77,* 1-13.

Gentile, R. M., Martino D. L., and Entz M. H. (2005) Influence of perennial forages on subsoil organic carbon in a long term rotation study in Uruguay. *Agriculture, Ecosystems and Environment, 105,* 419-423.

Gijsman, A. J., and Thomas, R. J. (1996) Evaluation of some physical properties of an oxisol after conversion of native Savanna into legume-based on pure grass pastures. *Tropical Grassland, 30,* 237-248.

Glinski, J., and Lipiec, J. (1990) Soil Physical Conditions and Plant Roots. (pp 250) CRC Press Inc, Boca Raton, Florida.

Grant, C. D., and Dexter, A. R. (1989) Generation of microcracks in moulded soils by rapid wetting. *Australian Journal of Agricultural Research, 27,* 169-182.

Grant, C. D., and Dexter, A. R. (1990) Air entrapment and differential swelling as factors in the mellowing of moulded soil during rapid wetting. *Australian Journal of Soil Research, 28,* 361-369.

Greenland, D. J. (1981) Soil Management and soil degradation. *Journal of Soil Science, 32,* 301-322.

Greenwood, K. L., MacLeod, D. A., Scott, J. M., and Hutchinson, K. J. (1997) Long-term stocking rate effects on soil physical properties. *Australian Journal of Experimental Agriculture, 37,* 413-419.

Greenwood, K. L., MacLeod, D. A., Scott, J. M., Hutchinson, K. J. (1998) Changes to soil physical properties grazing exclusion. *Soil Use and Management, 14,* 19-24.

Greenwood, K. L., and McKenzie, B. M. (2001) Grazing effects on soil physical properties and the consequences for pastures: a review. *Australian Journal of Experimental Agriculture, 41,* 1231-1250.

Gupta, S. C., and Allmaras, R. R. (1987) Models to assess the susceptibility of soils to excessive compaction. *Advances in Soil Science, 6,* 65-100.

Hamza, M. A., and Anderson W. K. (2003) Responses of soil properties and grain yields to deep ripping and gypsum application in a compacted loamy sand contrasted with a sandy clay loam soil in western Australia. *Australian Journal of Agricultural Research, 54,* 273-282.

Hamza, M. A., and Anderson W. K. (2005) Improving soil physical fertility and crop yield on a clay soil in western Australia. *Australian Journal of Soil Research, 53,* 615-620.

Hall, A. J., Rebella, C. M., Ghersa, C., and Culot, J. P. (1992) Field-crop systems of the Pampas. In C. J. Pearson (Ed.), *Field crop ecosystems.* (pp. 413–450) Elsevier, Amsterdam.

Haines, W. B. (1923) The volume changes associated with variations of water contents in soil. *Journal of Agricultural Science, 13,* 296-310.

Hartemink, A. E. (2002) Soil science in tropical and temperate regions. *Advances in Agronomy, 77,* 269-291.

Haynes, R. J., Swift, R. S., and Stephen, R. C. (1991) Influence of mixed cropping rotations (pasture-arable) on organic matter content, water stable aggregation and clod porosity in a group of soils. *Soil Tillage Research, 19,* 77-87.

Hoeppner, J. W., Entz, M. H., McConkey, B. G., Zentner, R. P., and Nagy, C. N. (2006) Energy use and efficiency in two Canadian organic and conventional crop production systems. Renew. *Agricultural and Food System, 21*, 60-67.

Holt, J. A., Bristol, K. L., and McIvor, J. G. (1996) The effects of grazing pressure on soil animals and hydraulic properties of two soils in semi-arid tropical Queensland. *Australian Journal of Soil Research, 34,* 69-79.

Imhoff, S., Pires Da Silva, A., and Tormena, C. A. (2000) Spatial heterogeneity of soil properties in areas under elephant-grass short-duration grazing system. *Plant and Soil, 219,* 161-168.

Lal, R. (2001) Managing world soils for food security and environmental quality. *Advances in Agronomy, 74,* 155-192.

Lal, R., Reicosky, D. C., and Hanson, J. D. (2007) Editorial: Evolution of the plow over 10,000 years and the rationale for no-till farming. *Soil and Tillage Research* 93: 1-12.

Larson, W. E., and Allmaras, R. R. (1971) Management factors and natural forces as related to compaction. In K. K. Barnes (Ed), *Compaction of Agricultural Soils.* American Society of Agricultural Engineers (pp. 367-427). St. Joseph, Michigan.

Mapfumo, E., Chanasyk, D. S., Naeth, M. A., and Baron, V. S. (1999) Soil compaction under grazing of annual and perennial forages. *Canadian Journal of Soil Science, 79,* 191-199.

Martens, D. A., Emmerich, W., McLain, J. E. T., and Johnsen, T. N. (2005) Atmospheric carbon mitigation potential of agricultural management in the southwestern USA. *Soil and Tillage Research, 83,* 95-119.

Martínez, L. J., and Zinck, J. A. (2004) Temporal variation of soil compaction and deterioration of soil quality in pasture areas of Colombian Amazonia. *Soil and Tillage Research, 75,* 3-17.

McCalla, G. R., Blackburn, W. H., and Merrill, L. B. (1984) Effects of livestock grazing on infiltration rates, Edwards Plateau of Texas. *Journal of Range Management, 37,* 265-269.

McGarry, D., and Daniells, I. G. (1987) Shrinkage curve indices to quantify cultivation effects on soil structure of a vertisol. *Soil Science Society of America Journal, 51,* 1575-1580.

Micucci, F. G., and Taboada, M. A. (2006) Soil physical properties and soybean (*Glycine max*, Merrill) root abundance in conventionally- and zero-tilled soils in the humid Pampa of Argentina. *Soil and Tillage Research, 86,* 152-162.

Miglierina, A. M., Iglesias, J. O., Landriscini, M. R., Galantini, J. A., and Rosell, R. A. (2000) The effects of crops rotation and fertilization on wheat productivity in the Pampean semiarid region of Argentina. 1. Soil physical and chemical properties. *Soil and Tillage Research, 53,* 129-135.

Monaghan, R. M., Paton, R. J., Smith, L. C., Drewry, J. J., and Littlejohn, R. P. (2005) The impacts of nitrogen fertilisation and increased stocking rate on pasture yield, soil physical condition and nutrient losses in drainage from a cattle-grazed pasture. *New Zealand Journal of Agricultural Research, 48,* 227-240.

Quiroga, A., Buschiazzo, D. E., and Peineman, N. (1999) Soil compaction is related to management practices in the semi-arid Argentine Pampas. *Soil and Tillage Research, 52,* 21-28.

Quiroga, A., Fernández, R., and Noellemeyer, E. (2009) Grazing effect on soil properties in conventional and no-till systems. *Soil and Tillage Research, 105,* 164-170.

Radford, B. J., Yule, D. F., McGarry, D., and Playford, C. (2001) Crop responses to applied soil compaction and to compaction repair treatments. *Soil and Tillage Research, 61,* 157-166.

Rodd, A. V., Papadoloulos, Y. A., Laflamne, L. F., McRae, K. B., Fillmore, S. A. E., and Wilson, R. W. (1999) Effect of rotational grazing on selected physical properties of a glayed Brunisolic Gray Luvisol loam in Nova Scotia. *Canadian Journal of Soil Science, 79,* 117-125.

Sadras, V. O., O'Leary G. J., and Rogeta D. K. (2005) Crop responses to compacted soil: capture and efficiency in the use of water and radiation. *Field Crops Research, 91,* 131-148.

Sanford, J. O. (1982) Straw and tillage management practices in soybean-wheat double cropping. *Agronomy Journal, 74,* 1032-1035.

Sasal, M. C., Andriulo, A. E., and Taboada, M. A. (2006) Soil porosity characteristics and water movement under zero tillage in silty soils in Argentinian Pampas. *Soil and Tillage Research, 87,* 9-18.

Satorre, E. H. (2004) Marco conceptual de la sustentabilidad: El suelo y la aplicación de pronósticos climáticos en la Región Pampeana. Seminario sustentabilidad de la producción agrícola. (pp. 2-5).

Senigagliesi, C., and Ferrari M. (1993) Soil and crop responses to alternative tillage practices. In D. R. Buxton, R. Shibles, R. A. Forsberg, B. L. Blad, K. H. Asay, G. M. Paulsen and R. F. Wilson (Eds), *Intenational crop Science I. Crop Science Society of America* (pp 27-35). Madison, Wisconsin, USA.

Singleton, P. L., and Addison, B. (1999) Effect of cattle treading on physical properties of three soils used for dairy farming in the Waikato, North Island. *Australian Journal of Soil Research, 37,* 891-902.

Singleton, P. L., Boyes, M., and Addison, A. (2000) Effect of treading by dairy cattle on topsoil physical conditions for six contrasting soil types in Waikato and Northland, New Zealand, with implications for monitoring. *New Zealand Journal of Agricultural Research, 43,* 559-567.

Siri-Prieto, G., and Ernst, O. (2010). Manejo Del suelo y rotación con pasturas: efecto sobre la calidad del suelo, el rendimiento de los cultivos y el uso de insumos. *Informaciones Agronómicas, 45,* 22-26.

Smil, V. (1999) Crop residues: Agriculture's largest harvest. *Bioscience, 49,* 299–308.

Smith, C. W., Johnston, M. A., and Lorentz, S. (1997) Assessing the compaction susceptibility of South African forestry soils. II. Soil properties affecting compactability and pressibility. *Soil and Tillage Research, 43,* 335-354.

Soriano, A., León, R. J. C., Sala, O. E., Lavado, R. S., Deregibus, V. A.,. Cauhépé, M. A, Scaglia, O. A., Velázquez, C. A., and Lemcoff J. H. (1991) Río de la Plata grasslands. In R. T. Coupland (Ed.), *Ecosystems of the world.* (pp. 367–407). Elseiver, Amsterdam.

Steinbach, H. S., and Alvarez, R. (2006) Changes in soil organic carbon contents and nitrous oxide emissions after introduction of no till in Pampean agroecosystems. *Journal of Environmental Quality, 35,* 3-13.

Stirk, G. B. (1954) Some aspects of soil shrinkage and the effect of cracking upon water entry into the soil. *Australian Journal of Agricultural Research, 5,* 279-290.

Strudley, M. W., Green, T. R., and Ascough, II J. C. (2008) Tillage effects on soil hydraulic properties in space and time: State of the science. *Soil and Tillage Research, 99,* 4–48.

Studdert, G. A., Echeverría, H. E., and Casanovas, E. M. (1997) Crop-pasture rotation for sustaining the quality and productivity of a Typic Argiudoll. *Soil Science Society of America Journal, 61,* 1466-1472.

Taboada, M. A., Micucci, F. G., Cosentino, D. J., and Lavado, R. S. (1998) Comparison of compaction induced by conventional and zero tillage in two soils of the rolling Pampa of Argentina. *Soil and Tillage Research, 49,* 57-63.

Taddese, G., Saleem, M. A. M., and Ayalneh, W. (2002) Effect of livestock grazing on physical properties of a cracking and self-mulching Vertisol. *Australian Journal of Experimental Agriculture, 42,* 129-133.

Tennant, D., and Hall, D. (2001) Improving water use of annual crops and pastures-limitations and opportunities in Western Australia. *Australian Journal of Agricultural Research, 52,* 171-182.

Thomas, G. W., Hazsler, G. R., and Robins, R. L. (1996) The effects of organic matter and tillage on maximum compactibility of soils using the Proctor test. *Soil Science, 161,* 502-508.

Thomas, G. A., Titmarsh, G.W, Freebairn, D. M., and Radford, B. J. (2007) No-tillage and conservation farming practices in grain growing areas of Queensland: A review of 40 years of development. *Australian Journal of Experimental Agriculture, 47,* 897–898.

Tisdall, J. M., and Oades, J. M. (1982) Organic matter-stable aggregates in soils. *Journal of Soil Science, 33,* 141-163.

Twerdoff, D. A., Chanasyk D. S., Mapfumo E., Naeth M. A., and Baron V. S. (1999) Impacts of forage grazing and cultivation on near-surface relative compaction. *Canadian Journal of Soil Science, 79,* 465-471.

Viglizzo, E. F. (1986) Agroecosystems stability in the argentine pampas. *Agriculture, Ecosystems and Environment, 16,* 1-12.

Viglizzo, E. F., and Frank, F. C. (2006) Land-use options for Del Plata Basin in South America: Tradeoffs analysis based on ecosystem service provision. *Ecological Economics, 57,* 140-151.

Willatt, S. T., and Pullar, D. M. (1983) Changes in soil physical properties under grazed pastures. *Australian Journal of Soil Research, 22,* 343-348.

Watson, R., and Zakri, A. H. (2005) MA Conceptual Framework. In R. Watson and A. H. Zakri (Eds.), *Ecosystems and Human Well-Being: Scenarios* (pp. 22-33). Island Press, Washington DC.

Zegwaard, K. E., Balks, M. R., and Singleton, P. L. (1998) The effects of cattle trampling and pugging on selected soil properties. *Proceedings of the New Zealand Soil Society Conference*, 101-102.

In: Grazing Ecology: Vegetation and Soil Impact
Editor: José Ramón Arévalo
ISBN 978-1-62100-436-3

Chapter 10

EFFECTS OF LIVESTOCK GRAZING ON OVERLAND FLOW AND SEDIMENT YIELD ON HILLSLOPES IN THE NORTHERN NEGEV, ISRAEL

Sarah Pariente*

Laboratory of Geomorphology and Soil, Department of Geography and Environment, Bar-Ilan University, Ramat-Gan, Israel

ABSTRACT

A mosaic-like spatial pattern, consisting of three main surface cover types (patches) – shrubs, trampling routes, and interspersed open areas – characterizes the landscape of the northern Negev region of Israel. This area has been grazed by flocks of sheep and goats since prehistoric times, and the presence of livestock has created a network of trampling routes which account for over 21% of the surface area of the landscape. The soil characteristics and hydrological responses of these routes differ from those of the remainder of the intershrub area and define a separate microhabitat. Each patch creates specific microenvironmental and functional conditions, which affect the eco-geomorphic system, i.e., relations among soil, water, plants and erosion, therefore, patches may differ from one another in their hydrological characteristics.

The impacts of grazing flocks on the distributions of overland flow and sediment yield, and on soil properties on south- and north-facing hillslopes in a semi-arid area were investigated. Overland flows and sediment yields, and also soil properties, were measured in small runoff plots (0.25–1.0 m^2) representing the three patch types, inside and outside 15-year-old long-term grazing exclosures.

Grazing reduces the spatial heterogeneity of overland flow and sediment yield, increases their total amounts and intensifies the source/sink relationships.

Keywords: overland flow, sediment yield, source/sink patches, soil properties, spatial distribution

* Email: pariens@mail.biu.ac.il

INTRODUCTION

Semi-arid environments are characterized by a non-uniform ground surface cover that comprises various components, such as vegetation, bare soil, exposed bedrock, rock fragments and soil crusts. Each component constitutes a separate micro-environment with its own pedological, hydrological, ecological and geomorphic behavior. Because of the effects of these varied characteristics on infiltration capacity, each micro-environment responds in its own manner to rainfall, and thereby affects the generation of overland flow and subsequent soil erosion. By these processes, water, sediments, litter and nutrients are redistributed on hillslopes, especially from bare ground areas to vegetated ones (Yair and Lavee, 1985; Lavee *et al.,* 1998; Li and Sarah, 2003). These areas function as sources and sinks, respectively (Bergkamp, 1998; Lavee *et al.,* 1998; Calvo-Cases *et al.,* 2003).

Landscape eco-geomorphic processes are driven by both natural forces and human activities. The latter have a wide impact on the direction and rate of processes that affect and are affected by hillslope characteristics such as ground cover and soil properties, as well as their complex interactions and feedbacks (Sarah, 2002). In semi-arid Mediterranean environments these anthropogenic influences have been present since prehistoric times, and livestock grazing is one of the commonest (Noy-Meir and Seligman, 1979).

In a survey of hydrological and erosional data obtained under natural or simulated rainfall on limestone landscapes in various Mediterranean environments in the last 15 years, Calvo-Cases *et al.* (2003) found that although several different environmental factors, e.g., climate, lithology, etc., were addressed, only a few studies took grazing into account.

Furthermore, conclusions about the redistribution of runoff and sediment yield in general, and the effects of grazing in particular (Manzano and Navar, 2000; Osem *et al.,* 2002; Golodets and Boeken, 2006; Stavi *et al.,* 2008a) were based on theoretical deductions derived from pedohydrological properties such as hydraulic conductivity (Bromley *et al.,* 1997), infiltration capacity (Eldridge *et al.,* 2000), soil moisture (Sarah, 2002; Katra *et al.,* 2007), soil organic contents (Sarah 2006), penetration resistance (Manzano and Navar, 2000; Stavi *et al.,* 2008a), vegetation properties (Puigdefabregas *et al.,* 1998; Shachak *et al.,* 1998; Golodets and Boeken, 2006) or simulation models (Ludwig *et al.,* 1999; Puigdefabregas *et al.,* 1999).

This chapter focuses on the impacts of grazing by sheep and goats on the spatial distribution of the actual overland flow and sediment yield in semi-arid areas in the northern Negev of Israel; before the special conditions of these rangelands are addressed, general aspects of landscape patchiness and its effect on the ecogeomorphic processes in semi-arid areas are summarized.

SOURCE/SINK RELATIONSHIPS IN SEMI-ARID ENVIRONMENTS

In semi-arid environments, the interactions between the high spatial variability of rainfall and local variations in landscape conditions and soil properties produce a complex matrix of source/sink micro-environments (Bergkamp, 1998; Puigdefabregas *et al.,* 1999), among which surface runoff, soil particles and nutrients are redistributed along the hillslopes. Many authors have suggested that the shrub patches function as trapping/sink areas for runoff and sediment (Yair and Lavee, 1985; Lavee *et al.,* 1998; Calvo-Cases *et al.,* 2003), whereas patches of bare soil, exposed bedrock, rock fragments (Arnau-Rosalén *et al.,* 2008), some biological soil crusts (Eldridge *et al.,* 2000), and livestock trampling routes (Stavi *et al.,* 2008b) function as sources for water and resources (Yair and Lavee, 1985; Lavee *et al.,* 1998; Calvo-Cases *et al.,* 2003). Organic matter and soil moisture contents, and soil stability (Sarah, 2002, 2006) and arylsulfatase activity (Zwikel *et al.,* 2007) were found to be greater, and sodium absorption ratio lower (Sarah, 2004) under shrubs than between shrubs. These findings indicate that shrubs act as water-collecting/sink areas and the bare soil serves as a water-contributing/source area (Lavee *et al.,* 1998).

This pattern of vegetation and bare surfaces, termed 'two-phase mosaic' by many authors (e.g., Ludwig *et al.,* 1999), has been shown to influence the infiltration characteristics of semi-arid hillslopes significantly (Abrahams and Parsons, 1991), because in these environments vegetation can persist only if water is added by overland flow from bare ground areas (Puigdefabregas and Sanchez, 1996). Consequently, patchiness is a cause of a self-organized spatial pattern of vegetation, that maximizes runoff harvest and minimizes sediment and nutrient loss (Puigdefabregas *and* Sanchez, 1996; Rietkerk *et al.,* 2002; Calvo-Cases *et al.,* 2003), and thereby preserves water and soil resources within the system (Tongway *and* Ludwig, 2003).

Runoff Generation and Sediment Movement in Semi-Arid Environments

In semi-arid environments, the non-uniform vegetation cover and the relative abundance of bare ground, rock outcrops, rock-fragment cover and soil crusts affect the infiltration capacity of the soils, and this, in turn, influences the generation of overland flow (Arnau-Rosalén *et al.,* 2008). In places where vegetation is scarce the soil is exposed to direct raindrop impact, which leads to rainsplash and crust formation (Rostango *and* del Valle, 1988). As a consequence the infiltration rate decreases exponentially with time during a rain event, and when rainfall exceeds the infiltration rate, overland flow is initiated. This process is remarkably influenced by rainfall intensity and begins after a relatively short period of time during the rain event. It is termed 'infiltration excess' or 'Hortonian' overland flow (Yair *and* Lavee, 1985; Lavee *et al.,* 1998). A second mechanism for generation of overland flow, which involves saturation of the topmost subsurface soil layer, is activated during rain events of relatively long duration, specifically after heavier amounts of rainfall; it is termed 'saturation excess' overland flow (Lavee *et al.,* 1998; Calvo-Cases *et al.,* 2003). This mechanism generally acts in areas where surface factors enhance infiltration and prevent rainsplash and crust formation, and consequently inhibit the initiation of hortonian overland flow. These factors comprise the presence of continuous vegetation cover, plant litter, some biological crusts, small stones lying on the soil surface, and soil structure and composition (Lavee *et al.,* 1998). The generation of saturation excess overland flow is also affected by lithology and soil type and depth, and it occurs mostly on the lower parts of hillslopes (close to the channel), where the water table is generally high because of subsurface flow, or in topographic depressions. Puigdefabregas et al. (1998, 1999) found that two mechanisms of saturation excess overland flow developed under Mediterranean arid conditions in soils on mica-schist bedrock. The first one depends on saturation of the uppermost soil layer just after rainfall initiation; the second one requires the persistence of nearly saturated conditions in a shallow sub-surface soil layer.

When the conditions for saturation excess overland flow are reached, they lead to a situation in which small additional amounts of rainfall may generate widespread overland flow that causes transfer of water and sediments on the hillslope scale (Puigdefabregas *et al.,* 1998, 1999). Under humid conditions, that are associated with prevalence of biotic factors, saturation excess overland flow is the primary means of redistribution of water and sediments along the hillslopes, whereas in arid environments, which are dominated by abiotic factors, infiltration excess overland flow is the primary mechanism (Lavee *et al.,* 1998; Imeson *et al.,* 1998). However, in semi-arid environments, under certain conditions overland flow can be driven by a combination of both processes (Calvo-Cases *et al.,* 2003).

Soil erosion can be caused by both water and wind action, but the present chapter addresses soil erosion caused by overland flow. Soil erosion depends greatly on the energy or detaching capacity of overland flow, in combination with the availability of free soil particles or the ease with which these particles are detached from their initial locations. (Yair and Lavee, 1981). Overland flow detaches and transports soil particles, litter, organic matter and nutrients, and whenever it reaches the channel and flows as runoff, resources leak out of the system (Shachak *et al.,* 1998).

THE EFFECTS OF VEGETATION ON OVERLAND FLOW GENERATION AND EROSION

Soil surface cover type is of crucial importance in determining the conditions for development of overland flow, because it has a very strong influence on infiltration capacity. Thus, vegetation clusters enhance the infiltration rate (Rostango and del Valle., 1988) by several means: (a) they present a physical barrier that moderates overland flow velocity and continuity (Bergkamp, 1998; Shachak *et al.,* 1998), traps soil particles and litter (Sanchez *and* Puigdefabregas, 1994), and furthers the development of shrub mounds (Rostango *and* del Valle., 1988; Parsons *et al.,* 1992), thereby changing the surface microtopography, soil texture and bulk density (Stavi *et al.,* 2008a, 2009); (b) shrubs soften direct raindrop impact, thereby enhancing infiltration capacity and preventing formation of soil crusts (Rostango and del Valle., 1988); (c) the combined physical, chemical and biological effects of shrub roots (Archer *et al.,* 2002) and soil biological activity (of micro- and macrofauna) (Garner and Steinberger, 1989) improve soil structure and create macropores in which water flows vertically at relatively high rates (Bromley *et al.,* 1997); and (d) there is significantly less trampling by livestock beneath the shrubs than on bare ground (Golodets and Boeken, 2006) and, consequently, less soil compaction.

All these factors and processes create a positive feedback mechanism (Calvo-Cases *et al.,* 2003): the enhanced infiltration beneath shrub patches minimizes generation of overland flow under their canopies and thereby strengthens the sink/source mechanism by which overland flow generated in the bare-soil patches is rapidly infiltrated into the soil beneath the shrubs (Bergkamp, 1998; Lavee *et al.,* 1998; Sarah, 2002), where it deposits soil particles, litter, nutrients and organic matter which, in turn, enhance the infiltration rate under the shrubs by changing the microtopography (Stavi *et al.,* 2008a) and improving soil conditions and properties (texture, structure, porosity, fertility and chemical composition) (Calvo-Cases *et al.,* 2003). Thus, shrubs act as traps for water and nutrients, and prevent leakage of these resources from the system, thereby maintaining its functionality (Shachak *et al.,* 1998; Tongway and Ludwig, 2003).

Grazing Effects on Runoff and Sediment Yield

Many processes that regulate landscape form and properties are greatly affected by present and past land uses (Calvo-Cases *et al.,* 2003). The main effects of grazing in semi-arid environments spring from the combined effects of herbivory and trampling on the one hand (Golodets and Boeken, 2006), and nutrient addition in the forms of urine and dung, on the other hand (McNaughton, 1979).

The combined engineering effects of herbivory and trampling affect the eco-geomorphic features of the landscape (Golodets and Boeken, 2006). Herbivory leads to a decrease in canopy and herbaceous cover, i.e., an increase in bare soil. Trampling leads to soil compaction, with increased soil density (Schlesinger *et al.,* 1990; Stavi *et al.,* 2008a) and destruction of topsoil structure (Manzano and Navar, 2000), especially in flock trampling routes (Stavi *et al.,* 2008b). These processes enhance subsequent rainsplash with consequently increased soil crusting, which eventually reduces soil infiltration and enhances surface runoff and soil erosion (Manzano and Navar, 2000; Sarah, 2002). Moreover, the decrease in vegetation cover leads to a decrease in the continuity of overland flow, thereby affecting the balance between source and sink micro-environments, and making the environment prone to erosion by water and wind (Shachak *et al.,* 1998; Schlesinger *et al.,* 1990).

Furthermore, grazing affects the redistribution of resources along the hillslope through several processes: (a) sporadic trampling in the intershrub bare soil patches detaches soil particles, rendering them prone to wind and overland flow erosion ; (b) browsing triggers a decrease in plant litter cover, which, in turn, decreases the nutrient exchange rate; and (c) the decrease in vegetation cover changes: (i) the radiation budget, and (ii) the soil moisture budget, as evaporation from the bare soil increases (Sarah and Rodeh, 2004).

As explained above, these processes have a positive feedback effect, which determines the present-day ecogeomorphic characteristics of the landscape, so that a relatively small shift in the direction or the rates of the processes involved can lead to considerable ecogeomorphic effects; hence the importance of the present chapter.

Grazing in the Northern Negev Region of Israel: Impact on the Pedohydrological Characteristics

The Lehavim Bedouin Demonstration Farm, is located in the northern Negev region of Israel (31°20' N, 34°46' E; Figure 1). This is a hilly, semi-arid area, 350-500 m above sea level, with mean annual precipitation of about 300 mm that falls between October/December and May (Perevolotsky and Landau, 1988; Bar'am, 1996; Osem *et al.,* 2002). Average daily temperature ranges from 10° C in the wet winter, to 25° C in the dry summer (Bar'am, 1996), and relative humidity ranges from 51% in May to 68% in January (Bitan and Rubin, 1991).

The lithology is chalk and limestone of the Eocene, and the soil type is Brown Rendzina. The soil is shallow, generally not deeper than 20 cm in open spaces between shrubs and 40 cm under shrubs, except in rock fissures. The color is dark brown (7.5YR4/3) and the texture is clay-loamy in which the primary particle-size distribution is 30% clay, 40% silt and 30% sand. The dominant clay type is montmorillonite, and the stone content is about 15–30% (Zwikel *et al.,* 2007).

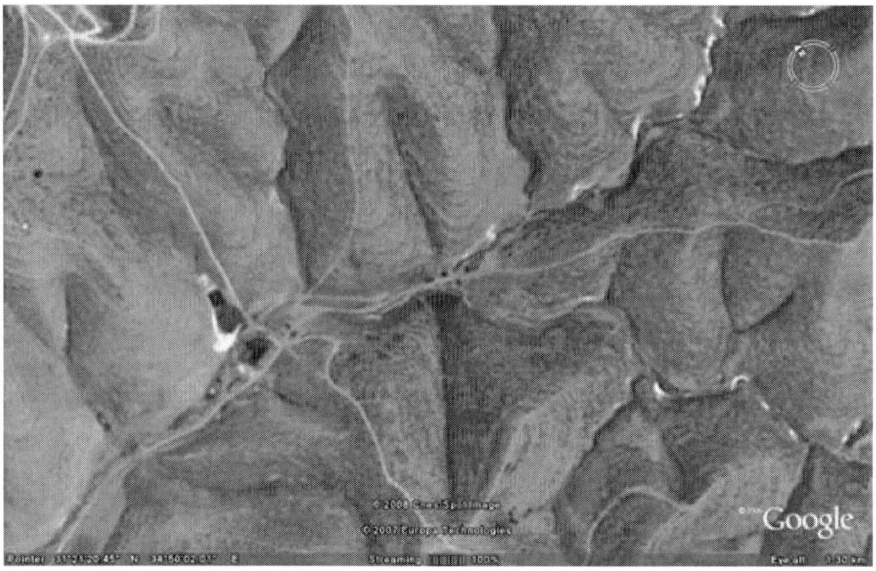

Figure 1. Study site: Lehavim Bedouin Demonstration Farm, northern Negev, Israel.

The mean gradient of the hillslopes is 15°. The vegetation physiognomy is sparse shrubland, with a patchy distribution of vegetation, biological crusts, exposed bedrock and bare soil (Osem *et al.,* 2002).

Vegetation is characterized by (a) dwarf shrubs, mainly *Sarcopoterium spinosum, Coridothymus capitatus* and *Thymelaea hirsute*; (b) perennial grasses and forbs, mainly *Asphodelus ramosus* and *Poa bulbosa*; and (c) annual herbaceous vegetation (Osem *et al.,* 2002).

On the hillslopes, the components forming the most extensive cover are herbaceous vegetation, trampling routes and shrubs (Stavi *et al.,* 2008b).

The area, like many other semi-arid areas of the Old World, has been grazed by flocks of sheep and goats since prehistoric times, i.e., 5,000–8,000 years, therefore the vegetation comprises mainly grazing-tolerant species (Noy-Meir and Seligman, 1979). The 800-ha farm was established in 1980 and since then has been managed under the auspices of Israel's Ministry of Agriculture and Rural Development; it is grazed by flocks of sheep and goats, totaling about 800 head of breeding stock (Figure 2).

Figure 2. Flock of sheep and goats.

METHODS

The simplest subdivision of the types of surface cover of many semi-arid rangelands is into a two-component mosaic of shrub patches interspersed with open spaces, in which the former act as sinks for water and other resources, and the latter as sources. However, close observations in the northern Negev area suggested that the spatial pattern of surface components is more complicated, and that the open space between shrubs consists of two components: herbaceous areas, separated by trampling routes that support no vegetation. Therefore, the apparent trampling routes are considered to function differently from the remainder of the intershrub area, with regard to pedo-hydrological characteristics, and are designated as a third type of surface cover (Figure 3).

Surface Cover Survey

A survey of the surface cover components of the area was conducted in the winter. Three plots measuring 4 × 4.5 m were randomly selected on each of two opposing north- and south-facing hillslopes, of slope 15 and 13°, respectively, in an area frequented by the flocks and subjected to intense grazing pressure. A metal frame measuring 2.0 × 1.3 m with wires stretched across it was constructed to form a grid of 260 cells, each measuring 0.1 × 0.1 m. The frame was supported by four telescopic legs that enabled it to stand horizontally. The area of each plot, apart from two central 0.3-m wide access strips, was mapped by means of six placements of the frame. For each cell the cover component that occupied the largest surface area was recorded (Stavi *et al.*, 2008b).

Figure 3. Dominant surface cover types in Lehavim Bedouin Demonstration Farm, northern Negev, Israel.

Soil – Field and Laboratory Work

Soil sampling was conducted on north- and south-facing hillslopes, at the peak of the growing season (March) and at the end of the hot, dry season (September).

On each hillslope three sampling areas in the middle of the slope were delineated. Each sampling area comprised one ungrazed control exclosure (established in 1994) and an adjacent grazing area, i.e., a total of 12 experimental plots. Soil was sampled from each of the three types of cover, which were chosen according to their dominance, as revealed by the surface cover survey: shrub (SH) – represented by *S. spinosum*, which had high coverage on both aspects; flock trampling route (RU); and the rest of the intershrub area (IS). For each type of cover, soil was sampled from seven randomly selected points within each of the 12 experimental plots. At each point soil samples were collected from depth ranges of 0–2 cm and 5–10 cm.

For each soil sample, moisture content, organic carbon content, bulk density, calcium carbonate content, and aggregate size distribution were determined.

Overland Flow and Sediment Yield Measurements

On the above hillslopes in an open area exposed to grazing and in a fenced area from which grazing had been excluded (control area), small runoff plots, 0.25–1.0 m^2 in area, were constructed. These plots were of three types, representing each of the above dominant types of ground surface cover. There were three or four replicates of each type of runoff plot.

The plots were bounded by concrete walls, 3–5 cm high, nailed to the ground and embedded 2 cm into the soil. A pipe was attached to the topographically lower end of each plot, from which water was collected into a bucket embedded in the ground.

The amount of rain was determined by small rain gauges mounted perpendicularly to the slope.

After each rain event, the buckets containing overland flow were replaced in the field with new ones, and were taken to the laboratory for determination of the gross weight of the overland flow and sediment yield.

Overland flow and sediments were collected during two consecutive winters: 2007/8 and 2008/9.

The total quantities of overland flow and sediment yield were determined according to weight, on the assumption that 1 cc of water weighs 1 g. The sediment yield was calculated by oven-drying the overland flow for 24 h at 105° C and weighing the sediments left in the vessel. The sediment yield weight (g) was then subtracted from the total weight of runoff water plus sediment, to determine the runoff volume (cc). Statistical analyses were applied with EXCEL and SAS software. The mean values of the various variables were compared among the micro-environments/treatments.

RESULTS AND DISCUSSION

Ground Surface Cover

The dominant shrub species on both aspects was *Sarcopoterium spinosum*, which is a moderately palatable cushion-like shrub, 20-60 cm in height, and with canopy width ranging from 30 cm for small individuals to more than 100 cm for older, well-developed patches. The shrub *Coridothymus capitatus* – an unpalatable cushion-like shrub, 30-40 cm in height, 30-50 cm in width – occurred only on the south-facing aspect. A litter layer about 1 cm thick was found underneath the canopy of both shrub species. The most dominant component of the intershrub area was herbaceous vegetation (primarily annual grasses and forbs), which had a very similar cover to that of *S. spinosum*: about 35 and 23% on the northern and southern hillslopes, respectively. The next most dominant component of the intershrub area comprised flock trampling routes of almost bare, compacted soil. These traversed the hillslopes with a predominantly horizontal orientation, and were visually recognizable because of their mechanically formed crusts and very sparse herbaceous cover. The cover of these routes was similar on both aspects and accounted for an average of 21% of the area (Stavi *et al.*, 2008b).

Soil Properties

Mean values of soil properties under the three cover types, in the control and grazed areas, are presented in Figure 4. Regarding the effect of cover type on soil properties, Figure 4 shows that the soil moisture and organic carbon contents were highest under the shrubs and lowest under the trampling routes, and had intermediate values under the intershrub areas. Bulk density and calcium carbonate content were lowest under the shrubs and highest under the trampling routes, with intermediate values under the intershrub areas. A comparison between the grazed and control areas showed that soil moisture and organic matter contents under the shrubs and in the intershrub patches was higher in the grazed than in the control areas, whereas this order was reversed under the trampling routes. In the grazed areas, bulk density under the trampling routes was significantly higher than that in the control. Calcium carbonate content under the shrub was significantly higher in the grazed area than in the control.

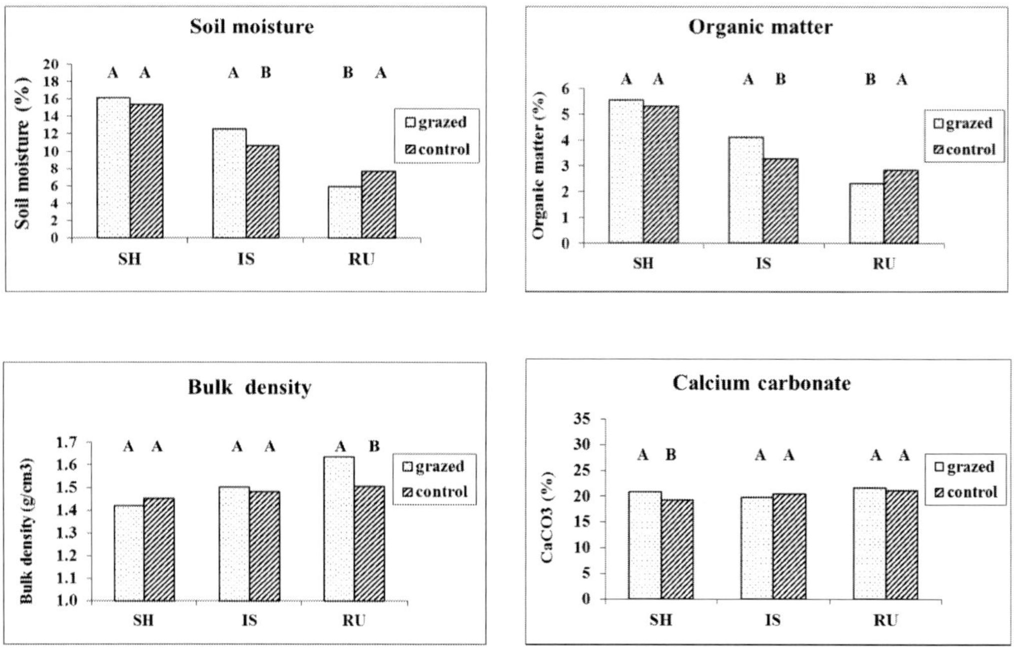

Figure 4. Soil properties in grazed and control areas (treatments) for the various surface cover types, in 0-2-cm soil depth, in the spring. The values represent both the northern and southern hillslopes. For each surface cover type, different letters indicate significant ($p < 0.05$) differences between the grazed and control areas.

Livestock affects soil properties via two main mechanisms: trampling compacts the soil and increases its bulk density (Van Haveren, 1983; Warren *et al.,* 1986; Trimble and Mendel, 1995); and herbivory reduces plant cover (Weltz and Wood, 1986; Coughenour, 1991).

Grazing increased bulk density and reduced moisture and organic carbon contents but, almost invariably, only along the trampling routes, where there was less vegetation cover than elsewhere. In contrast, in the shrub patches, which exhibited the highest permanent vegetation cover, and where this cover persisted year-round, grazing did not affect the soil bulk density, nor its moisture and organic matter contents. Also, the bulk density under shrubs was unaffected by trampling, which contrasts with the findings of Golodets and Boeken (2006) in a different site in the northern Negev region: they found a significant effect of sheep trampling on soil compaction of soil mounds under the shrub *Noaea mucronata*. This difference is attributed to differences in canopy architecture between *S. spinosum* and *N. mucronata*, and also to differences between the studies: in grazing intensity, in animal herd composition, and in microtopography. Browsing might increase losses of soil water via increased evaporation, and might reduce litter input which would lead to depletion in soil moisture and organic matter contents. Although field observations indicated that sheep and goats browsed the foliage of *S. spinosum*, there was no significant effect of grazing on the moisture and organic matter contents of the soil under the shrub. This highlights the great importance of the shrub patches in maintaining sustainable ecosystems in arid rangelands. Differences between the grazed and control areas in runoff yield from the shrub patch might account for the significant difference in soil calcium carbonate content. Shrubs in the grazed area contribute larger amounts of runoff than those in the control (see next section), so that more of the calcareous dust that reaches the soil is washed out from the shrub patches in the former area.

The higher soil moisture content under the intershrub patches in the grazed area than in the ungrazed one is attributed to decreased transpiration in the grazed area because herbivory reduced the plant cover. In addition, the trampling routes contributed more runoff water to their downslope areas in the grazed than in the ungrazed area (see next section). A similar phenomenon was observed in the semi-arid Judean Desert of Israel by Sarah and Lavee (personal communication), who found that the soil moisture content of grazed intershrub areas was higher than that of ungrazed ones during the growing season.

Soil organic matter content under the intershrub patches was also greater in the grazed than in the ungrazed areas. This reflects the higher soil moisture content, which enhanced primary production and, in turn, litter production and decomposition (Raiesi and Asadi, 2006; Yan *et al.*, 2007), with consequent increase in the soil organic matter content.

The bulk density of the soil under the intershrub patches did not differ between the grazed and ungrazed areas, because a major part of the trampling load was concentrated along the trampling routes.

Prevention of grazing for over a decade reduced the contrast in the ground surface features and soil properties between the former trampling routes and the remainder of the intershrub area, but did not eliminate this contrast. In the control exclosures, former trampling routes were much less recognizable because of the almost complete disappearance of the mechanical crusts and the greater herbaceous cover than on the existing trampling routes. Soil bulk density in the upper soil layer under the trampling routes was significantly lower inside the exclosure than outside it, which resulted in increased infiltration capacity and moisture content of soil under the trampling routes inside the exclosure. Such favorable soil conditions and litter input encourages seed germination and enhances herbaceous colonization and, in turn, litter production and decomposition; it thereby led to a significantly higher soil organic content under the former trampling routes than under the grazed areas, within the exclosures.

Overland Flow and Sediment Yield

Mean values of specific runoff, i.e., the volume of runoff yield per square meter, and specific sediment yield, i.e., the weight of sediment yield per square meter, in grazed and ungrazed (control) areas on south-facing hillslopes are presented in Figures 5 and 6. The north-facing hillslope showed similar trends in runoff and sediment yield among the various cover types to those on the south-facing one, but smaller absolute amounts of these variables, therefore the present chapter relates only to the latter. The results show that in the control areas both runoff and sediment yield were affected by soil surface cover type: they increased from the shrub patches, through the intershrub patches, to the trampling routes. Whereas the shrub patch produced the lowest specific runoff (0.4 mm) and sediment yield (0.3 g/m^2), the former trampling routes produced the most: 4.8 mm and 5.2 g/m^2, respectively. In the remainder of the intershrub patches, intermediate values were found: 2.21 mm and 2.3 g/m^2. The coefficients of variation of the runoff and of the sediment yield were 90 and 95%, respectively. In the grazed areas the shrub patch prduced the lowest specific runoff and sediment yield, 1.21 mm and 1.1 g/m^2, respectively, the trampling routes produced 7.4 mm and 16.0 g/m^2, and the remainder of the intershrub patches, 7.4 mm and 25.7 g/m^2. Coefficients of variation of the runoff and the sediment yield were 67% and 87%, respectively.

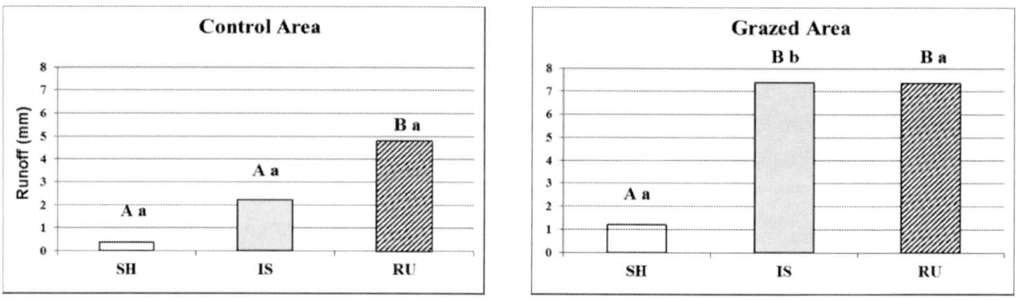

Figure 5. Runoff yield in the various surface cover types in grazed and control areas. Southern hillslope, winter 2007/8. Different capital letters indicate significant ($p < 0.05$) differences between the surface cover types within a treatment. Different small letters indicate significant ($p < 0.05$) differences for each surface cover type between treatments.

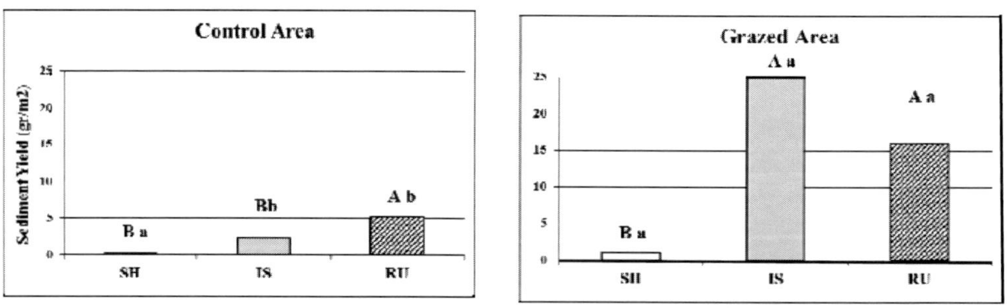

Figure 6. Sediment yield in the various surface cover types in grazed and control areas. Southern hillslope, winter 2007/8. Different capital letters indicate significant ($p < 0.05$) differences between the surface cover types within a treatment. Different small letters indicate significant ($p < 0.05$) differences for each surface cover type between treatments.

Runoff and sediment yield, for each surface cover type and for each hillslope aspect, were higher in the grazed than in the ungrazed areas. However, the amounts of the above variables were notably greater on the southern hillslope.

The soil surface cover types differed in their hydrological responses to rainfall. The specific runoff yield increased from shrub patches, through intershrub patches to trampling routes. This runoff production sequence also was associated with surface vegetation cover: the less vegetation cover the greater the runoff yield, for the reasons described in section "The effect of vegetation on overland flow...". Whereas shrubs provided the highest vegetation cover, with consequently the greatest capability to trap water and nutrients – i.e., to function as sinks – and to prevent leakage of these resources from the hillslope system, the trampling routes carried the least cover, and therefore acted as sources. The latter consideration applied also, but to a lesser extent, to the intershrub patches. These results are consistent with those of other studies that suggested that the shrub patches functioned as sinks for runoff and sediment, whereas patches of bare soil and livestock trampling routes functioned as sources of water and resources (Yair and Lavee, 1985; Lavee *et al.,* 1998; Calvo-Cases *et al.,* 2003; Stavi *et al.,* 2008a).

In the control un-grazed areas, the absolute differences in runoff yield between shrub and intershrub patches, between the intershrubs patches and former trampling route, and between shrub patches and former route were 1.8, 2.6 and 4.4 mm, respectively. The absolute differences in sediments yield between shrub and intershrubs patches, between intershrub patches and former trampling route, and between shrub patches and former route were 2.1, 2.8 and 4.9 g/m^2, respectively.

In the grazed areas, the absolute differences in runoff and sediment between shrub and intershrubs patches, between intershrub patches and trampling route, and between shrub patches and route were 6.2, 0.0 and 6.2 mm, and 24.6, 9.7 and 14.9 g/m^2, respectively.

Grazing increased runoff and sediment yield on all surface cover types. In the shrub, intershrub and route patches grazing increased runoff by 3.1, 3.4 and 1.5 times, and sediment yields by 4.1, 11 and 3.1 times, respectively. The finding of the lowest increases in the route patches was attributed to the persistence of high runoff and sediment generation on the former routes, which had experienced intense animal trampling that resulted in high soil compaction in the past; i.e., the soil in the routes still "remembers" its land use history, which it expressed in slow recovery of this microenvironment.

Under grazing the route and intershrub microenvironments had similar outputs of overland flow but that these differed from that of the shrub microenvironment. This means that grazing shifted the spatial distribution of runoff towards lower spatial heterogeneity, but intensified the spatial differences with respect to areas that were not subjected to this activity. However, the CVs in the grazed and control areas indicate that grazing reduced the spatial heterogeneity of sediment yields to a lesser extent than that of runoff. This means that the occurrence of grazing influenced the source/sink relations through modification of the landscape mosaic.

The effect of grazing on the runoff/erosion relationships in the grazed area was more pronounced in the intershrub (IS) microenvironments than in the trampling route (RU) and shrub (SH) ones. In the IS microenvironment soil erosion was more affected by runoff than in the RU and SH ones: in the three microenvironments each millimeter of runoff eroded 3.4, 2.2, and 0.9 g/m^2 of sediments, respectively. Shrubs reduce soil erosion by runoff because of the combined effects of consolidation of soil aggregates – by roots, and high organic matter and soil moisture contents – and the low energy of the runoff because of the flattened microtopography. The trampling routes reduce the erosive effect of runoff because their soil is the most compacted; the soil particles are organized in a dense structure that reduces vulnerability to shear strength. These two soil cover types exhibited self-regulation with respect to soil erosion whereas the intershrub areas did not, except when they were covered with herbaceous vegetation towards the spring, and the increase in herbaceous vegetation caused the infiltration rate to increase and runoff to decrease.

Conclusions

A hierarchical structure exists in the semi-arid ecogeomorphic system in the Negev: the soil properties (first layer) are associated with the vegetation cover types (second layer), and these, affect runoff and sediment yield (third layer). In rangelands this "web" pattern encourages regulation of the redistribution of resources on the hillslopes by changing the surface cover type.

The presence of livestock created a network of trampling routes which account for over 21% of the surface area of the landscape. The soil characteristics and hydrological response under these routes differ from those under the remainder of the intershrub patches; they thereby increase the small-scale fragmentation of the ecological pattern and constitute a separate microhabitat. This means that the trampling routes play an important role in ecosystem functioning.

Grazing reduces the spatial heterogeneity of overland flow and sediment yield, increases their total amounts and intensifies the source/sink relationships.

The above non-trophic effects are readily subsumed within the definition of an ecosystem engineer: an organism that regulates the productivity of other organisms by controlling their resource supply or by modifying their habitat (Jones *et al.,* 1997).

Because the impact of grazing – through its creation of source areas, overland flow and erosion – is higher on the southern than on the northern aspect, the former hillslopes are more fragile than the latter.

Clearance of the shrubs, as a management practice, will cause the formation of gradual homogeneity of the ground surface and soil properties, leading to a greater continuity of overland flow and soil erosion at the hillslope scale. This means a decrease in the functionality of the ecosystem.

Acknowledgment

I am grateful to Hanoch Lavee, Eugene David Ungar, Ilan Stavi, Moy Zonana, Nathan Fragin, Helena Zhevelev, Eyal Sachs and Dani Meir, for their helpful assistance in the field and laboratory.

References

Abrahams, A. D., and Parsons, A. J. (1991) Relation between infiltration and stone cover on a semiarid hillslope, southern Arizona. *Journal of Hydrology*, 122, 49-59.

Archer, N. A., Quinton L. J. N., and Hess, T. M. (2002) Below-ground relationships of soil texture, roots and hydraulic conductivity in two-phase mosaic vegetation in south-east Spain. *Journal of Arid Environments*, 52, 535-553.

Arnau-Rosalén, E., Calvo-Cases, A., Boix-Fayos, C., Lavee, H., and Sarah, P. (2008) Analysis of soil surface component patterns affecting runoff generation. An example of methods applied to Mediterranean hillslopes in Alicante (Spain). *Geomorphology*, 101, 595–606.

Bar'am, H., 1996. *Meteorological data. Lehavim Hills, Israel (1987–95)*. Volcani Institute, Bet Dagan, Israel. (in Hebrew).

Bergkamp, G. (1998) A hierarchical view of the interactions of runoff and infiltration with vegetation and micro-topography in semiarid shrublands. *Catena*, 33, 201–220.

Bitan, A., and Rubin, S. (1991) *Climatic atlas of Israel for physical and environmental planning and design*. Department of Geography, Tel Aviv University, Tel Aviv, and Israel Meteorological Service, Ministry of Transport, Bet Dagan.

Bromley, J., Brouwer, J., Barker, A. P., Gaze, S. R., and Valentin, C. (1997) The role of surface water redistribution in an area of patterned vegetation in a semi-arid environment, south-west Niger. *Journal of Hydrology*, 198, 1–29.

Calvo-Cases, A., Boix-Fayos, C., and Imeson, A. C. (2003) Runoff generation, sediment movement and soil water behavior on calcareous (limestone) slopes of some Mediterranean environments in southeast Spain. *Geomorphology*, 50, 269–291.

Coughenour, M. B. (1991) Spatial components of plant-herbivore interactions in pastoral, ranching, and native ungulate ecosystems. *Journal of Range Management*, 44, 530–542.

Eldridge, D. J., Zaady, E., and Shachak, M. (2000) Infiltration through three contrasting biological soil crusts in patterned landscapes in the Negev, Israel. *Catena*, 40, 323-336.

Garner, W., and Steinberger, Y. (1989) A proposed mechanism for the formation of 'fertile islands' in the desert ecosystem. *Journal of Arid Environments*, 16, 257–262.

Golodets, C., and Boeken, B. (2006) Moderate sheep grazing in semiarid shrubland alters small-scale soil surface structure and patch properties. *Catena*, 65, 285-291.

Imeson, A. C., Lavee, H., Calvo A., and Cerda, A. (1998) The erosional response of calcareous soils along a climatological gradient in Southeast Spain. *Geomorphology*, 24, 3–16.

Jones, C. G., Lawton, J. H., and Shachak, M. (1997) Positive and negative effects of organisms as physical ecosystem engineers. *Ecology*, 78, 1946–1957.

Katra, I., Blumberg, D. G., Lavee, H., and Sarah, P. (2007) Top soil moisture patterns on arid hillsides—micro-scale mapping by thermal infrared images. *Journal of Hydrology*, 334, 359–367.

Lavee, H., Imeson, A. C., and Sarah, P. (1998) The impact of climate change on geomorphology and desertification along a Mediterranean-arid transect. *Land Degradation and Development*, 9, 407–422.

Li, X., and Sarah, P. (2003) Arylsulfatase activity of soil microbial biomass along a Mediterranean – arid transect. *Soil Biology and Biochemistry*, 35, 925-934.

Ludwig, J. A., Tongway, D. J., and Marsden, S. G. (1999) Stripes, strands or stipples: modeling the influence of three landscape banding patterns on resource capture and productivity in semi-arid woodlands, Australia. *Catena*, 37, 257–273.

Manzano, M. G., and Navar, J. (2000) Processes of desertification by goats overgrazing in the Tamaulipan thornscrub (matorral) in north-eastern Mexico. *Journal of Arid Environments*, 44, 1-17.

McNaughton, S. J. (1979) Grazing as an optimization process: grass-ungulate relationships in the Serengeti. *The American Naturalist*, 113, 691-703.

Noy-Meir, I., and Seligman, N. (1979) Management of semi-arid ecosystems in Israel. In B. H. Walker (ed.) *Management of semi-arid ecosystems* (pp. 113–160). Amsterdam, Netherlands: Elsevier.

Osem, Y., Perevolotsky, A., and Kigel, J. (2002) Grazing effect on diversity of annual plant communities in a semi-arid rangeland: interactions with small-scale spatial and temporal variation in primary productivity. *Journal of Ecology*, 90, 936-946.

Parsons, A. J., Abrahams, A. D., and Simanton, J. R. (1992) Microtopography and soil-surface materials on semi-arid piedmont hillslopes, southern Arizona. *Journal of Arid Environments*, 22, 107–115.

Perevolotsky, A., and Landau, Y. (1988) *Improvement and development of the Bedouin stock industry in the northern Negev, Lehavim demonstration farm. Professional report 1982-1988*. Department of Natural Resources, Institute of Field and Garden Crops, Agricultural Research Organization, Bet Dagan, Israel. (in Hebrew).

Puigdefabregas, J., and Sanchez, G. (1996) Geomorphological implications of vegetation patchiness in semi-arid slopes. In: M. Anderson, and S. Brooks (Eds.), *Advances in hillslope processes* (pp. 1027–1060). London: Wiley.

Puigdefabregas, J., del Barrio, G., Boer, M., Gutierrez, L., and Sole, A. (1998) Differential responses of hillslope and channel elements to rainfall events in a semi-arid area. *Geomorphology*, 23, 337–351.

Puigdefábregas, J., Solé, A., Gutiérrez, L., del Barrio, G., and Boer, M. (1999) Scales and processes of water and sediment redistribution in drylands: results from the Rambla Honda field site in Southeast Spain. *Earth-Science Reviews*, 48, 39–70.

Raiesi, F. and E. Asadi. (2006) Soil microbial activity and litter turnover in native grazed and ungrazed rangelands in a semi-arid ecosystem. *Biology and Fertiity of Soils*, 43, 76-82.

Rietkerk, M., Boerlijst, M. C., van Langevelde, F., Hille Ris Lambers, R., van de Koppel, J., Kumar, L., Prins, H. H. T., and de Roos, A. M. (2002) Self-organization of vegetation in arid ecosystems. *American Naturalist*, 160, 524–530.

Rostango, C. M., and del Valle, H. F. (1988) Mounds associated with shrubs in aridic soils of northeastern Patagonia: characteristics and probable genesis. *Catena*, 15, 347–359.

Sanchez, G., and Puigdefabregas, J. (1994) Interaction between plant growth and sediment movement in semi-arid slopes. *Geomorphology*, 9, 243–260.

Sarah, P. (2002) Spatial patterns of soil moisture as affected by shrubs, in different climatic conditions. *Environmental Monitoring and Assessment*, 73, 237–241.

Sarah, P. (2004) Soil sodium and potassium adsorption ratio along a Mediterranean-arid transect. *Journal of Arid Environments*, 59, 731–741.

Sarah, P. (2006) Soil organic matter and land degradation in semi-arid area, Israel. *Catena*, 67, 50–55.

Sarah, P., and Rodeh, Y. (2004) Soil structure variations under water and vegetation manipulations. *Journal of Arid Environments*, 58, 43–57.

Schlesinger, W. H., Reynolds, J. F., Cunningham, G. L., Huenneke, L. F., Jarrell, W. M., Virginia, R. A., and Whitford, W. G. (1990) Biological feedbacks in global desertification. *Science*, 247, 1043–1048.

Shachak, M., Sachs, M., and Moshe, I. (1998) Ecosystem management of desertified shrublands in Israel. *Ecosystems*, 1, 475–483.

Stavi, I. Ungar, U., Lavee, H., and Sarah, P. (2008a) Surface microtopography and soil penetration resistance associated with shrub patches in a semiarid rangeland. *Geomorphology*, 94, 69–78.

Stavi, I. Ungar, U., Lavee, H., and Sarah, P. (2008b) Grazing-induced spatial variability of soil bulk density and content of moisture, organic carbon and calcium carbonate in a semi-arid rangeland. *Catena*, 75, 288–296.

Stavi, I., Ungar, U., Lavee, H., and Sarah, P. (2009) Livestock modify ground surface microtopography and penetration resistance in a semi-arid shrubland. *Arid Land Research and Management*, 23, 237–247.

Tongway, D. J., and Ludwig, J. A. (2003) The nature of landscape dysfunction in rangelands. In: J. A. Ludwig, D. J. Tongway, D. Freudenberger, J. Noble, and K. Hodgkinson (Eds.), *Landscape ecology function and management* (pp. 49–61). Canberra: CSIRO Publishing.

Trimble, S. W., and Mendel, A. C. (1995) The cow as geomorphic agent – a critical review. *Geomorphology*, 13, 233–253.

Van Haveren, B. P. (1983) Soil bulk density as influenced by grazing intensity and soil type on a short grass prairie site. *Journal of Range Management*, 36, 586–588.

Warren, S. D., Thurow, T. L., Blackburn, W. H., and Garza, N. E. (1986) The influence of livestock trampling under intensive rotation grazing on soil hydrologic characteristics. *Journal of Range Management*, 39, 491–495.

Weltz, M., and Wood, M. K. (1986) Short duration grazing in central New Mexico: effects on infiltration rates. *Journal of Range Management*, 39, 365–368.

Yair, A., and Lavee, H. (1981) An investigation of source areas of sediment and sediment transport by overland flow along arid hillslopes. Erosion and sediment transport measurement. *IAHS Publication*, 133, 433–446.

Yair, A., and Lavee, H. (1985) Runoff generation in arid and semi-arid zones. In M. G. Anderson, and T. P. Burt (Eds.), *Hydrological forecasting* (pp. 183–220). New York: John Wiley and Sons.

Yan, H., Cao, M., Liu, J., and Tao, B. (2007) Potential and sustainability for carbon sequestration with improved soil management in agricultural soils of China. *Agriculture Ecosystems and Environment*, 121, 325–335.

Zwikel, S., Lavee, H., and Sarah, P. (2007) Temporal dynamics of arylsulfatase enzyme activity in various microenvironments along a climatic transect in Israel. *Geoderma*, 140, 30–41.

INDEX

D

E

I

J

K

L

M

S